ENERGY SCIENCE, ENGINEERING AND TECHNOLOGY

CELLULOSIC BIOFUELS: FEEDSTOCK, TECHNOLOGY AND POLICY OPTIONS

ENERGY SCIENCE, ENGINEERING AND TECHNOLOGY

Additional books in this series can be found on Nova's website under the Series tab.

Additional E-books in this series can be found on Nova's website under the E-books tab.

AGRICULTURE ISSUES AND POLICIES

Additional books in this series can be found on Nova's website under the Series tab.

Additional E-books in this series can be found on Nova's website under the E-books tab.

ENERGY SCIENCE, ENGINEERING AND TECHNOLOGY

CELLULOSIC BIOFUELS: FEEDSTOCK, TECHNOLOGY AND POLICY OPTIONS

ROBERT M. BURKIN
AND
ALEX C. RILEY
EDITORS

Nova Science Publishers, Inc.
New York

For permission to use material from this book please contact us:
Telephone 631-231-7269; Fax 631-231-8175
Web Site: http://www.novapublishers.com

LIBRARY OF CONGRESS CATALOGING-IN-PUBLICATION DATA

Cellulosic biofuels : feedstock, technology and policy options / editors:
Robert M. Burkin and Alex C. Riley.
p. cm.
Includes index.
ISBN 978-1-61209-509-7 (hardcover)
1. Biomass energy--Government policy--United States. I. Burkin, Robert M.
II. Riley, Alex C.
HD9502.5.B543U623 2011
333.95'390973--dc22
2010051520

Published by Nova Science Publishers, Inc. † New York

CONTENTS

PREFACE

Cellulosic biofuels are thought by many to hold the key to increased benefits from renewable biofuels because they are made from potentially low-cost, diverse, non-food feedstocks. They could also potentially decrease the fossil energy required to produce ethanol, resulting in lower greenhouse gas emissions. This new book summarizes the current state of knowledge regarding potential biomass feedstocks, production and marketing constraints, processing technologies and the economics of biomass from field to fuel under current and hypothetical policy circumstances.

Chapter 1- In the Energy Independence and Security Act of 2007 (P.L. 110-140), Congress mandated the use of a large and rapidly increasing volume of biofuels as part of the U.S. national transportation fuel base. In particular, the share of cellulosic biofuels is mandated to grow to 16 billion gallons by 2022—a daunting challenge considering that no commercial production existed as of mid-2010. Cellulosic biofuels can be produced from almost any sort of biomass. As a result, a variety of biomass types that can be produced or collected under a range of geographic settings are potential feedstock sources. However, part of the mandate's challenge will be encouraging farmers to produce or collect non-traditional biomass materials that require multiple growing seasons to become established, and for which markets currently do not exist. Participation represents a substantial risk for producers, and even under the most optimistic conditions, U.S. agriculture will be challenged to produce the enormous volume of biomass needed to meet the biofuels mandate.

Chapter 2- Cellulosic biofuels are produced from cellulose (fibrous material) derived from renewable biomass. They are thought by many to hold the key to increased benefits from renewable biofuels because they are made from potentially low-cost, diverse, non-food feedstocks. Cellulosic biofuels could also potentially decrease the fossil energy required to produce ethanol, resulting in lower greenhouse gas emissions.

Chapter 3- The Renewable Fuel Standard (RFS) was expanded under the Energy Independence and Security Act of 2007 (EISA; P.L. 110-140) in an effort to reduce dependence on foreign oil, promote biofuel use, and stabilize transportation fuel prices, among other goals. Over a 15-year period, the RFS seeks to establish a market for biofuels in the transportation sector by requiring that increasing amounts of biofuels-36 billion gallons by 2022—be blended into transportation fuel. The mandate is to be accomplished with an assortment of advanced biofuels, including cellulosic biofuels—fuels produced from cellulosic materials including grasses, trees, and agricultural and municipal wastes. The cellulosic biofuel allotment in the mandate, as established by Congress in EISA, was 100 million gallons due in 2010, increasing to 16 billion gallons by 2022. However, on March 26, 2010, the U.S. Environmental Protection Agency (EPA) issued a fmal rule for implementation of the RFS that sets a new, lower cellulosic biofuel mandate of 65 million gallons for 2010.

Chapter 4- One of our greatest challenges is to reduce our nation's dependence on imported petroleum. To accomplish this, we need a variety of alternative fuels, including ethanol produced from cellulosic materials like grasses and wood chips. Fortunately, the United States has abundant agricultural and forest resources that can be converted into biofuels. Recent studies by the U.S. Department of Energy (DOE) suggest these resources can be used to produce enough ethanol – 60 billion gallons/year – to displace about 30% of our current gasoline consumption by 2030.

In: Cellulosic Biofuels: Feedstock, Technology… ISBN: 978-1-61209-509-7
Editors: Robert M. Burkin et al.

Chapter 1

CELLULOSIC ETHANOL: FEEDSTOCKS, CONVERSION TECHNOLOGIES, ECONOMICS, AND POLICY OPTIONS

Randy Schnepf

SUMMARY

In the Energy Independence and Security Act of 2007 (P.L. 110-140), Congress mandated the use of a large and rapidly increasing volume of biofuels as part of the U.S. national transportation fuel base. In particular, the share of cellulosic biofuels is mandated to grow to 16 billion gallons by 2022—a daunting challenge considering that no commercial production existed as of mid-2010. Cellulosic biofuels can be produced from almost any sort of biomass. As a result, a variety of biomass types that can be produced or collected under a range of geographic settings are potential feedstock sources. However, part of the mandate's challenge will be encouraging farmers to produce or collect non-traditional biomass materials that require multiple growing seasons to become established, and for which markets currently do not exist. Participation represents a substantial risk for producers, and even under the most optimistic conditions, U.S. agriculture will be challenged to produce the enormous volume of biomass needed to meet the biofuels mandate.

Potential biomass feedstocks are numerous and widespread throughout the United States, and include woody biomass, perennial grasses, and agricultural and forest residues. Each type of biomass faces tradeoffs in terms of production, storage, and transportation. Dedicated energy and tree crops have large up-front establishment costs and will likely take several years to produce a commercial harvest, but can produce high yields with relatively low maintenance costs thereafter. Residues are nearly costless to produce, but confront difficult collection strategies and do not always produce uniform biomass for processing. Agricultural residues face complicated trade-offs between soil nutrient loss and biomass yield, as well as questions about the optimal timing strategy for harvesting the main crop and residue (either jointly or separately). Logging residues confront a tradeoff with energy production at the plant (via burning).

None of the potential feedstocks (other than starch from corn) are economical to convert into biofuels under current commercial technology without substantial federal policy intervention. In addition to federal policy and the choice of feedstock, the processing technology used, the distribution infrastructure, and blending rates are expected to play major roles in the economic viability of cellulosic biofuels. Different processing technologies yield different biofuels in terms of energy content and usability, while also strongly influencing the economic viability of biofuels production. Ethanol produced under current biochemical processes yields only 67% of the energy of an equivalent volume of gasoline, and (due to its chemical properties) cannot use the same storage tanks, pipelines, and retail pumps as gasoline. In contrast, synthetic petroleum products (i.e., green hydrocarbons) obtained from biomass processed using more costly thermochemical technology yield an energy content nearly equal to petroleum fuels and can be used in existing fuel infrastructure. Currently ethanol is blended in most gasoline at about a 10% rate. If the rising usage mandate is to be met, the biofuels blending rate will necessarily have to increase, at which point the energy equivalence of a biofuel will likely influence the choice of processing technology, distribution infrastructure, and federal policy incentives.

Many uncertainties remain concerning biomass producer participation rates, the choice of biomass, and associated yields and costs of production, harvest, storage, and transportation, as well as contractual marketing arrangements, plant location, and conversion technology, among other issues. This chapter attempts to summarize the current state of knowledge regarding potential biomass feedstocks, production and marketing constraints, processing technologies, and the economics of biomass from field to fuel under current and hypothetical policy circumstances. As such, it is intended to serve as a reference for policymakers interested in understanding the complexity underlying the development of a large-scale, biomass-based fuel system.

An executive summary of the report is available in Chapter 1.

INTRODUCTION

Under the Energy Independence and Security Act of 2007 (EISA; P.L. 110-140), Congress mandated the use of a large and rapidly increasing volume of biofuels as part of the U.S. national transportation fuel base. In particular, the share of cellulosic biofuels is mandated to grow to 16 billion gallons by 2022—a daunting challenge considering that no commercial production existed as of mid-2010. In addition to the biofuels use mandate, Congress also provides federal support in the form of tax credits to fuel blenders and biofuels producers, and an import tariff on foreign- produced ethanol to protect and encourage the development of the U.S. biofuels industry.

Despite this strong federal support, many uncertainties remain over whether a large-scale, economically viable cellulosic biofuels system can be successfully developed. These uncertainties include the following.

- Which biomass source and processing technology will provide the highest energy yields at the lowest cost?
- What incentives will encourage biomass producers to cultivate dedicated crops that may take three or more years before they are established and produce marketable output, and for which no market presently exists?
- How will large volumes of biomass—that must be produced within a narrow temperate-zone harvest window (e.g., March to October)—be harvested, dried, stored, and ultimately transported to a processing plant that must operate throughout the year?
- What changes will be needed in the U.S. transportation-fuel infrastructure to facilitate distributing and consuming the mandated rapid, large expansion of cellulosic biofuels?
- What set of federal policies can best facilitate the development of such a system?

- Will there be regional consequences within the United States from the emergence of such a system? What about potential international consequences?

It is still too early to begin to answer many of the broader social welfare questions, such as who will be potential winners and losers in the development of a large-scale biomass-based biofuels system. However, substantial research has been done in recent years concerning the economics of production, harvest, and energy yield for various biomass sources under different processing technologies. A review of this research can help to clarify current bottlenecks in the development of a cellulosic biofuels industry and provide some guidance to policymakers looking to extend or modify existing federal policy, or formulate new policy in support of such a biofuels industry.

This chapter attempts to summarize the current state of knowledge regarding potential biomass feedstocks, production and marketing constraints, processing technologies, and the economics of biomass from field to fuel under current and hypothetical policy circumstances. As such, it is intended to serve as a reference for policymakers interested in understanding the complexity underlying the development of a large-scale, biomass-based fuel system.

This chapter provides a broad overview of the nascent U.S. cellulosic biofuels industry and the many uncertainties associated with its future. This assessment was conducted by a team of researchers at Purdue University's Department of Agricultural Economics. The report provides a "snapshot" of current technological development, but is both prospective and retrospective because it also examines emerging or advanced technologies that may affect future biofuels development, and looks at evidence from a growing body of research on the economics of biomass production and biofuels processing as guidelines for shaping energy policy.

STRUCTURE OF THE REPORT

The report contains five chapters. The first chapter is an Executive Summary, which provides an overview of the report's main findings. The Executive Summary is followed by the main report, which consists of Chapters 2-5, together with figures and tables listed in the Table of Contents. Each chapter can be read independently; however, Chapters 2-4

("Introduction," "Lingocellulosic Feedstocks," and "Cellulosic Biofuel Conversion Technologies") provide the reader with background and context for a more complete understanding of the economic analysis and discussion contained in the final chapter.

REPORT AUTHORSHIP

This technology assessment and report was written by Purdue University, Department of Agricultural Economics, under the leadership of Wallace E. Tyner,[1] together with Sarah Brechbill[2] and David Perkis.[3] The report's authorship rests with Tyner, Brechbill, and Perkis.[4] The work was performed under contract to CRS, and is part of a multiyear CRS project to examine different aspects of U.S. energy policy. This chapter was funded, in part, by a grant from the Joyce Foundation. Randy Schnepf served as the CRS project coordinator.

1. EXECUTIVE SUMMARY

Cellulosic Biofuels

Cellulose-based biofuels are thought to offer substantial advantages over current corn ethanol, foremost that they can be grown at low cost on marginal land where they will not compete with traditional food crops. In addition, cellulosic biomass sources are abundant and widely distributed throughout the United States.

Changes made in the U.S. Renewable Fuel Standard (RFS) in 2007 require renewable fuels use of 36 billion gallons by 2022, more than triple the nearly 11 billion gallons consumed in the United States in 2009. A substantial portion (16 billion gallons) of the 2022 RFS is required to come from cellulosic biofuels, an industry in its infancy and not yet economically viable. Considerable uncertainty remains about how a cellulosic biofuels industry will evolve, what infrastructure changes will be needed to support this development, and how the development of a biomass- based biofuels industry will alter regional economic and environmental circumstances within the United States.

The major objective of this chapter is to summarize what is and is not known about cellulosic biofuels. What do we know about the technical and economic potential of the various feedstocks and conversion processes? What are the key bottlenecks or impediments to the development of a cellulosic biofuels industry? What are the likely impacts of the government mandates and incentives related to cellulosic biofuels? This paper attempts to answer these questions based on published literature and model simulations.

The Choice of Feedstock

Current research on feedstocks focuses on maximizing yields, harvesting and collecting efficiently, and testing various supply chains to minimize losses and overall delivered costs.

Feedstock characteristics. The preferred biomass feedstocks, known as lignocellulosic feedstocks, are composed of three main parts: cellulose, hemicellulose, and lignin. The different concentrations of these components in particular feedstocks will affect the efficacy of the conversion technology. Feedstocks with larger quantities of cellulose and hemicellulose are favored in biochemical conversion processes as the conversion technology stands presently, but research is being conducted to incorporate lignin into conversion or modify some of its properties through hybrid feedstocks. As a result, research offers the potential to alter the viability and preference of various feedstocks.

Feedstock sources. Potential biomass feedstocks are numerous and widespread throughout the United States, and include agricultural residues from such crops as corn, wheat, rice, and sugarcane, perennial grasses such as switchgrass and *Miscanthus*, short-rotation woody crops such as poplar and willow, and forest residues removed directly from the forest or taken from mills after processing. In 2010, the Environmental Protection Agency estimated that the 16 billion gallons of cellulosic biofuels mandated by the RFS for 2022 will be derived from dedicated energy crops (49.4%), agricultural residues (35.6%), urban waste (14.4%), and forest residue (0.6%). Several studies based on agronomically viable U.S. biomass production suggest that the biofuels potential is significantly larger than 16 billion gallons.

Feedstock production and harvest costs. Each type of biomass faces tradeoffs in terms of production, storage, and transportation. Dedicated energy and tree crops have large up-front establishment costs and may take several years to produce a commercial harvest, but can produce high yields with relatively low maintenance costs thereafter. Residues are nearly costless to produce, but confront difficult collection strategies and do not always produce uniform biomass for processing. Agricultural residues also face complicated trade-offs between soil nutrient loss and biomass yield, as well as questions about the optimal timing strategy for harvesting the main crop and the residue (either jointly or separately). Logging residues confront both a tradeoff with energy production at the plant (via burning) and a uniformity of quality issue for processing.

Producer participation. A key aspect of biomass production will be understanding what economic incentives will encourage producer participation in growing and/or collecting biomass feedstocks. Whether the feedstock is a residue or dedicated crop, there is not much experience among producers, markets do not presently exist for most potential biomass crops, and few risk management tools (e.g., crop insurance) are available to producers. Studies presently use a wide range of estimated participation—30% to 80%—depending on the size and location of the processing plant, the price being paid for biomass by the plant, the cost of producing, harvesting, and collecting biomass for a producer of a given size, and the terms of producer contracts.

Feedstock transportation. From the perspective of the biofuels processing plant, the choice of which feedstock to use will depend on the location of the plant and the local feedstock supply availability. Agricultural residues and perennial grass feedstocks have relatively low energy density, which means a large volume would be needed by a plant to meet demand. As a result, transportation costs may be crucial in the biomass choice. For example, a 50-million-gallon-peryear cellulosic ethanol plant (that converts a ton of corn stover—which includes the stalks, leaves, and cobs of the corn plant—into 90 gallons of biofuel) would need over 550,000 tons of corn stover harvested from roughly 280,000 acres of land (assuming 2 tons per acre of removable corn stover) in order to stay operational for a year. Because trucks become physically full when hauling biomass before reaching their weight limits, more truckloads would be required for cellulosic biofuel production relative to corn-based ethanol, thus making the per-unit transport cost much higher for cellulosic feedstocks than for corn. The greater number of trucks required to transport cellulosic feedstocks might also impose a heavy burden on rural transportation infrastructure. Compacting the biomass prior to transport may result in net savings if transport cost savings offset densification costs.

Feedstock storage. Because the harvest of dedicated energy crops and agricultural residue is concentrated during one part of the year, long-term storage will be needed to hold the biomass until the plant is ready to use it (keeping in mind that biofuels plants are designed to run year- round). For plants that can eventually use multiple feedstocks, they will be able to use fall-harvested biomass around the time it is harvested, followed by perennial grasses with their wider harvest period, or woody biomass or urban waste that can become available throughout the year. Storage costs, inclusive of biomass dry-matter loss due to weathering and exposure, could be crucial in determining optimal feedstock sources in certain agro-climatic zones.

Uniformity of feedstock. Biomass can have highly inconsistent quality and characteristics, not only among feedstocks but within a feedstock type. Delivering a uniform feedstock to the plant will decrease processing and pretreatment costs. Plants receiving biomass that has been harvested differently and stored under a variety of conditions will incur additional costs to arrive at a uniform feedstock product. With more experience and the development of the appropriate harvesting, collecting, and storage technologies, both standardization and densification of biomass could take place before the biomass even arrives at the plant, which would lower per-unit transportation costs and allow the plant to start with a uniform and consistent product, regardless of its source.

The Choice of Processing Technology

Two primary biomass-to-biofuel conversion methods for lignocellulosic feedstocks are currently under consideration for commercial use—biochemical conversion and thermochemical conversion. Energy yields and production costs for each process may vary substantially based on the specific implementation design and feedstock.

Biochemical Conversion

Process. The biochemical conversion process is similar to the process currently used to produce ethanol from corn starch. Enzymes or acids are used to break down a plant's cellulose into sugars, which are then fermented into liquid fuel. Four key steps are involved. First, feedstock is pretreated by changing its chemical makeup to separate the cellulose and hemicellulose from the lignin in order to maximize the amount of available sugar. Second, hydrolysis uses enzymes or acids to break down the complex chains of sugar molecules into simple sugars for fermentation. Third, fermentation is used to convert the sugar into liquid fuel. Fourth, the liquid fuel is distilled to achieve a 95% pure form.

Production costs. Each step of the conversion process incurs costs. A key cost of *pretreatment* is the time incurred. Biological pretreatment, using fungi, for example, can take 10 to 14 days. Chemical pretreatment may be faster but with higher costs for chemicals. *Hydrolysis* cost depends largely on the cost of enzymes, which has been estimated at $0.50 per gallon. However, hydrolysis may require different enzymes for different parts of the plant, thus incurring varying enzyme costs for different feedstocks. The effectiveness of hydrolysis is highly dependent on the effectiveness of pretreatment—too much lignin remaining after pretreatment will impede enzyme efficiency. More accessible sugar (following efficient pretreatment) may improve enzyme function at lower costs. Regarding *fermentation*, different sugar types, for example, hexoses (six- carbon sugars) or pentoses (five-carbon sugars), may require different yeast strains for fermentation. Improved yeast strains that ferment various sugars could lower production costs. Some cost savings may be available through management of *recovered lignin*, which can be burned to generate electricity and steam to power the bio-refinery or for other outside uses. Primary research areas for potential improvement in energy yield and cost reduction are in pretreatment, hydrolysis, and fermentation. Distillation is already a well-established technology. Successful improvements should make the process adaptable to multiple feedstocks.

Energy equivalency. The biochemical conversion of biomass into alcohol produces a liquid fuel (i.e., ethanol) that contains only about 67% of the energy content of gasoline. When ethanol is blended with gasoline at low rates (e.g., 10% or less), the reduced gas mileage resulting from the blended fuel is sufficiently small that most consumers are likely to treat the blended fuel as equivalent volumetrically (i.e., gallon for gallon) with gasoline. However, at higher blend rates, especially at an 85%-ethanol blend rate (E85), the lower mileage is more noticeable and consumers may prefer energy-equivalent pricing, whereby the price for a gallon of E85 should be only 72% of the price of a gallon of gasoline.

Petroleum infrastructure equivalency. Because of its physical properties, ethanol cannot be used in the same infrastructure (e.g., pipelines, storage tanks, service pumps) used to deliver retail gasoline. Nor can it be used directly by standard vehicle systems that have not been adjusted for ethanol blends greater than 10% (or 15% for model year 2007 or newer light-duty vehicles). This both limits ethanol retail delivery opportunities and raises the cost of delivery.

Thermochemical Conversion

Process. Thermochemical conversion processes, which use heat to decompose the feedstock, are well established and developed. Unlike biochemical conversion, thermochemical conversion uses the entire biomass, including the lignin portion. There are two main types of thermochemical conversion processes—gasification and pyrolysis.

Gasification is an anaerobic process where the partial combustion of biomass feedstock at over 700°C generates synthesis gas, or syngas, a mixture of carbon monoxide and hydrogen. The syngas must be cleaned of tar, ash, and other impurities prior to the next processing step. The Fischer-Tropsch process is then used to convert the "cleaned" syngas into a variety of liquid fuels. The presence of impurities in syngas might disrupt the Fischer-Tropsch process by inactivating the catalyst. Also, prior to gasification, the biomass must have 20% or less moisture. For practically any type of biomass, drying will be required. Established drying technologies usually take place at high temperatures, which creates an opportunity for improvement. Gasification is prohibitively expensive and would only be used commercially at extremely high gasoline prices.

Pyrolysis is the partial combustion of biomass feedstock at 450°C to 600°C in the presence of no oxygen, which produces bio-oil. Bio-oil, which is rich in carbon, is similar to crude oil and must be refined into biofuels. Fast pyrolysis requires higher temperatures than slow pyrolysis but occurs in about two seconds. Currently, fast pyrolysis is receiving the most attention as a viable conversion process. Keeping the pyrolysis oil stable long enough to transform the bio-oil into hydrocarbons is one of the major barriers in the pyrolysis pathway. Because pyrolysis converts the biomass into a liquid form, it is easier to store and transport.

Energy and infrastructure equivalency. The thermochemical conversion of biomass into synthetic fuels (or green hydrocarbons) produces liquid fuels that are essentially energy- equivalent to their petroleum counterparts, and fully adaptable for use in existing petroleum fuel infrastructure. As a result, energy-equivalent pricing favors thermochemically processed biofuels over biochemically processed biofuels.

Research keys. With respect to gasification, research is being done to dry biomass at lower temperatures and use excess heat from drying for other purposes. With respect to pyrolysis, stability of bio-oil and general cost reductions are the major research issues.

Commercialization Status

Neither conversion process is ready for commercialization. Presently, most cellulosic biofuels production is taking place in laboratories and small-scale demonstration and pilot plants. Plans for commercial plants have been announced by several companies, but development is likely to be slow, absent significant incentives. Substantial cost reductions will be necessary for successful commercial development. Several processes are currently being researched and developed in laboratories, but it is difficult to know with any certainty whether those that appear successful in trials will also be successful on a commercial scale. Given the lack of commercial production to date, per-gallon cost estimates for cellulosic biofuels are highly uncertain, but estimates based on laboratory and pilot plant results range from $2.50 per gallon to slightly over $3.00 per gallon.

The Department of Energy Biomass Program's Theoretical Ethanol Yield Calculator calculates the maximum theoretical biofuel yield per ton of feedstock with biochemical conversion of feedstocks based on their composition. Based on its results, the theoretical yields for corn stover, switchgrass, and forest thinning are 113, 97, and 82 gallons per dry ton, respectively. As plants and technologies approach commercialization, the rate of efficiency could approach 100% of the maximum theoretical energy yield.

Economic Comparison of Biomass, Processing Technology, and Policy Choices

Current biofuels production technologies (biochemical and thermochemical) were analyzed under the market conditions that prevailed during mid-2010, using a deterministic simulation model (with and without the tax credit subsidy).

Under *volumetric pricing*, with the current ethanol tax credit of $0.45 per gallon, the average corn-ethanol plant breaks even (operates at zero profit) when the price of oil is $56.33 per barrel (and the stochastically modeled corn price is $3.41 per bushel). In other words, the tax credit allows ethanol plants to operate profitably whenever oil prices are $56.33 per barrel or higher. Without the ethanol tax credit, the breakeven price of oil rises to $71.45 per barrel (still below recent market prices). Higher oil prices are needed to offset higher corn prices. The breakeven oil price scenarios are substantially higher for cellulosic biofuels—even with a much higher tax credit of $1.01 per gallon, whether produced from biochemical or thermochemical processes ($92.74 and $113.77, respectively, with tax credit).

Under *energy-equivalent pricing*, the breakeven oil price for corn ethanol rises considerably— $91.62 with the tax credit, and $114.19 without the tax credit—as the consumer has to spend more money to obtain the same energy (or mileage) available from petroleum fuels. The breakeven oil price is prohibitive for biochemical cellulosic biofuels ($145.98 with subsidy; $196.64 without). However, because its energy yield is nearly equal to fossil fuels, thermochemical biofuels appear more competitive ($98.92 with subsidy; $143.92 without).

Simulation results suggest that corn ethanol is profitable only if based on volumetric pricing as compared to energy-equivalent pricing. Biomass-based cellulosic biofuels are not yet profitable under either pricing approach (volumetric or energy-equivalent), even with a much higher tax credit of $1.01 per gallon. Currently, the *fixed tax credit* ($0.45 per gallon) incentivizes biofuels production during periods of low petroleum fuel prices; however, during periods of high oil prices (when biofuels are inherently more competitive with fossil fuels), the fixed subsidy adds to plant profitability at taxpayers' expense while encouraging greater biofuels production than is sought by the marketplace. A *variable subsidy* that declines with higher oil prices has been proposed as a policy tool to maintain biofuels incentives while limiting taxpayer exposure. This chapter includes a comparative analysis of a fixed versus variable tax credit, not as a policy recommendation, but strictly as a comparative analysis to aid Congress's understanding of the differences between the two policy options. Model results found that the variable subsidy lowered producer risk and taxpayer costs relative to the fixed subsidy, but had slightly lower average returns—largely because subsidy payments were not made at high oil prices.

2. INTRODUCTION

The major objective of this paper is to summarize what is and is not known about cellulosic biofuels. For years cellulose-based biofuels have been touted as the future of bioenergy. What is the situation today? What do we know about the technical and economic potential of the various feedstocks and conversion processes? What are the likely impacts of the government mandates and incentives related to cellulosic biofuels? This chapter attempts to answer these questions based on published literature and model simulations.

Recent changes in the Renewable Fuel Standard (RFS) in the United States will require renewable fuels production to more than triple in the next 12 years (by 2022), and those increases are expected to come largely from cellulosic biofuels, an industry in its infancy and not yet economically viable.[5] Cellulosic feedstocks of interest include agricultural residues from such crops as corn, wheat, rice, and sugarcane, perennial grasses such as switchgrass and *Miscanthus*, short-rotation woody crops such as poplar and willow, and forest residues removed directly from the forest or taken from mills after processing, just to name a few. Current research on feedstocks focuses on maximizing yields, harvesting and collecting efficiently, and testing various supply chains to minimize losses and overall delivered costs (**Figure 1**). Cellulosic conversion technologies are not the same as for corn ethanol, as this biomass material is much more complicated than starch-based feedstocks.[6] Conversion technologies receiving the most attention include the biochemical and the thermochemical approaches. These technologies are not completely new, but their previous applications differ greatly from cellulosic biofuels.

With both the supply of feedstocks and conversion technologies still in the early stages of development and neither having yet arrived at a commercial scale, future progress in both areas is uncertain. It remains to be seen whether the development of one will come to dominate the development of the other, or whether the two can evolve simultaneously while still arriving at a workable outcome to fulfill the advanced biofuels requirements of the RFS.[7] The primary theme for both feedstocks and conversion technologies is cost reduction in order to make commercialization a reality and to make cellulosic biofuels competitive with other energy sources. Research and development must make cellulosic biofuels production as flexible as possible in order to accommodate future improvements. Governments and universities will play an important role in reducing risk and providing guidance as feedstock producers and biofuels plants break ground on cellulosic biofuels production.

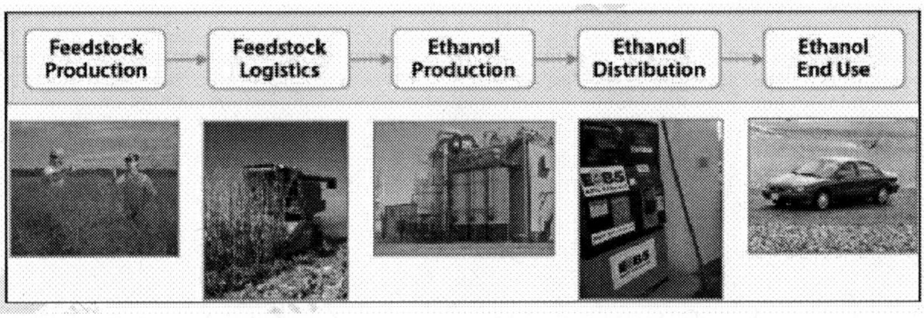

Source: U.S. Department of Energy, Energy Efficiency and Renewable Energy, Alternative Fuels and Advanced Vehicles Data Center, 2009, at *http://www.afdc.energy*

Figure 1. Cellulosic Ethanol Supply Chain

This chapter focuses on feedstock production, feedstock logistics, and biofuel production; however, challenges and obstacles to the distribution infrastructure and end use of biofuels in vehicles are also discussed. The remainder of the report is divided into three sections.

The first section ("Chapter 3: Lignocellulosic Feedstocks") covers the various types of lignocellulosic feedstocks, their respective characteristics, and a brief summary of how each feedstock is established and maintained. Then yields and production costs are outlined from a variety of studies to show the potential market effect of location, input costs, and assumptions. Next is a discussion of supply logistics and the challenges this new type of feedstock presents with respect to storage, quality, and transportation.

The second section of the report ("Chapter 4: Cellulosic Biofuel Conversion Technologies") considers biochemical and thermochemical conversion technologies. After a summary of each of the processes, the expected and theoretical energy yields of each technology are outlined. A summary of current estimates of capital and operating costs for each technology allows for discussion of where potential cost reductions might be made. Finally, a current list of planned and proposed pilot and demonstration plants is presented, with a discussion of how funding from the Department of Energy has been allocated so far.

The final section of the report ("Chapter 5: Economics and Policy of Cellulosic Biofuels") discusses the current status of relevant biofuels policy and some of the potential future policy options related to cellulosic biofuels. It also analyzes the effects of changes in these policies on the economics of cellulosic biofuel production. In particular, a simulation model is used to compare the market effects of a fixed subsidy (representative of the current fixed tax credit subsidy) and a variable subsidy that declines to zero as oil prices rise to a threshold level.

3. Lignocellulosic Feedstocks

Lignocellulosic feedstocks are composed of three main parts: cellulose, hemicellulose, and lignin. Depending on the conversion process used, the concentration of these components in a particular feedstock will affect the efficacy of biofuel production. Cellulose is a sugar polymer chain of glucose, or six carbon sugars. Hemicellulose is a sugar polymer chain of xylose, or five carbon sugars. Lignin forms the hard plant cell walls and cannot be fermented into liquid fuels in a biochemical conversion process, as can cellulose and hemicellulose. Lignin, however, can be utilized in the thermochemical conversion process and serves as a byproduct in the biochemical conversion process useful for providing energy to power the plant and even for generating electricity. The characteristics of the different parts of biomass and the amount of each present in a particular feedstock play a role in determining the efficacy of conversion technologies. The remainder of this chapter will discuss the production, yields, costs, and supply of the major lignocellulosic feedstocks in the United States—agricultural residues, perennial grasses, forest residues, and short-rotation woody crops.

Feedstock Types

Table 1 gives estimates of the composition of some feedstocks of interest. Feedstocks with larger quantities of cellulose and hemicelluloses are favored in biochemical conversion processes as the conversion technology stands presently, but research is being conducted to incorporate lignin into conversion or modify some of its properties through hybrid feedstocks. As a result, research offers the potential to alter the viability and preference of various feedstocks.

In addition to cellulose, hemicellulose, and lignin, lignocellulosic feedstocks also include organic acids, ash, proteins, oils, minerals, and other compounds.[8] **Figure 2** uses the composition estimates from **Table 1** to determine the average composition of corn stover, switchgrass, *Miscanthus*, and hardwoods. Examples of biomass resources include the following.

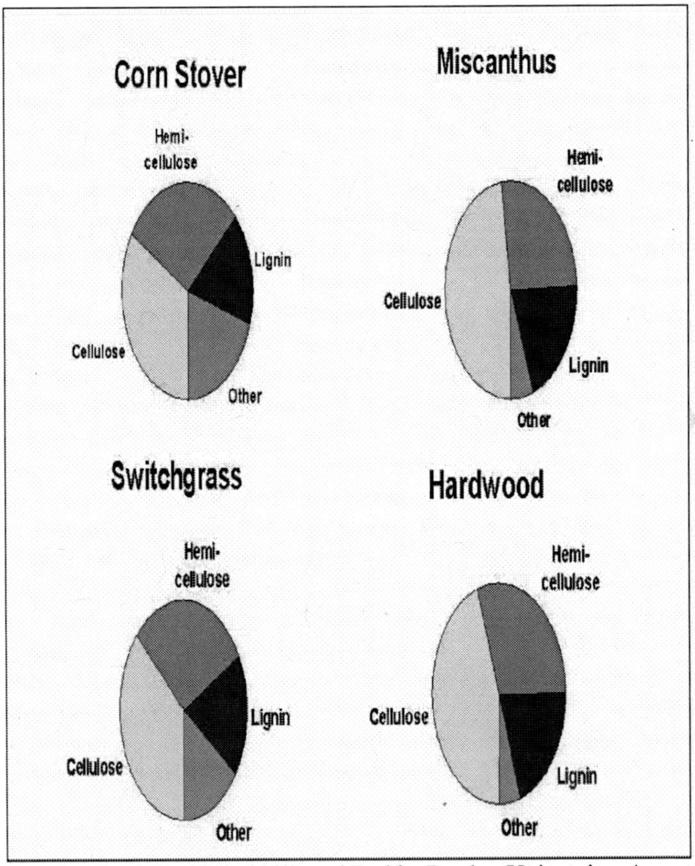

Source: Compiled by Tyner, Brechbill, and Perkis, Purdue University, August 2010.

Note: "Other" includes organic acids, ash, proteins, oils, minerals, and other compounds. Each pie chart is constructed from the simple average of the different estimates in Table 1.

Figure 2. Composition of Lignocellulosic Feedstocks by Type

Table 1. Estimated Composition of Lignocellulosic Feedstocks

Feedstock	Cellulose (%)	Hemicellulose (%)	Lignin (%)	Source (Year)
Corn stover	35	28	16-21	Scurlock (undated)[a]
	38	26	19	NC Sun Grant Center (2007)[b]
	32	44	13	Bull (1991)[c]
	31	19	18	Bransby (2007)[d]
Hardwood	45	30	20	Scurlock (undated)[a]
	38-50	25-35	15-25	Taylor (2009)[e]
	45	19	26	Hamelinck et al. (2003)[f]
Wheat straw	38	29	15	NC Sun Grant Center (2007)[b]
	38	36	16	Bull (1991)[c]
Softwood	42	21	26	Scurlock (undated)[a]
	50	23	22	Bull (1991)[c]
	41	18	28	Bransby (2007)[d]
Herbaceous energy crops	45	30	15	Bull (1991)[c]
Switchgrass	44-51	42-50	13-20	Scurlock (undated)[a]
	37	29	19	NC Sun Grant Center (2007)[b]
	31	24	18	Bransby (2007)[d]
	32	25	18	Hamelinck et al. (2003)[f]
Miscanthus	44	24	17	Scurlock (undated)[a]
	43	24	19	NC Sun Grant Center (2007)[b]

Source: Compiled by Tyner, Brechbill, and Perkis, Purdue University, August 2010, from sources listed.

[a] J. Scurlock, "Bioenergy Feedstock Characteristics," *Biomass Basics: Fact Sheets*, Oak Ridge National Laboratory (Oak Ridge, TN), undated.

[b] North Central Sun Grant Center, *Composition of Herbaceous Biomass Feedstocks*, Sun Grant Initiative—North Central Center, South Dakota State University (Brookings, SD), June 2007.

[c] S. R. Bull, U.S. Department of Energy, Biofuels Research Program, *Energy Sources, Part A: Recovery, Utilization, and Environmental Effects*, 13(4), 1991, pp. 443-442.

[d] D. I. Bransby, *Cellulosic Biofuel Technologies*, "Alternative Transportation Fuels Program," Alabama Department of Economic and Community Affairs (Montgomery, AL), February 2007.

[e] E. L. Taylor, "Co-products and By-products of Woody Biorefinery Processing," *Transition to a Bio Economy: The Role of Extension in Energy*, Farm Foundation Conference (Little Rock, AR), 2009.

[f] C. N. Hamelinck, G. van Hooijdonk, and A. P. C. Faaij, *Prospects for ethanol from lignocellulosic biomass: technoeconomic performance as development progresses*, Report NWS-E-2003-55, Utrecht University and Copernicus Institute (Utrecht, Netherlands), 2003.

- Agricultural crop residues (corn, wheat, rice, etc.)
- Perennial grasses (switchgrass, *Miscanthus*)
- Short-rotation woody crops (poplar, willow, eucalyptus)
- Conventional logging residues, wood processing mills residues, and removal of excess wood from forestlands
- Manure
- Food/feed processing residues
- Municipal solid waste and urban wood waste

Dedicated Energy Crops

Perennial grasses, including switchgrass and *Miscanthus*, have high yields (especially in warmer areas with a longer growing season), are low-maintenance, and have a more positive environmental impact than producing ethanol from corn. However, perennial grasses are not currently being grown on a widespread basis. Unlike residues left after the harvest of a primary commercial crop, grasses themselves are primary crops that require a relatively long-term commitment of resources to growing a crop that remains relatively obscure. As a primary crop used for energy production, and unlike residues used for energy production, perennial grasses are the only source of revenue on a given area of land. Mistakes with establishment and maintenance serve to deplete potential profits. There is no commodity market to guarantee the price of perennial grasses or to allow producers to sell their product freely. The development of contracts between producers and processing plants will prove particularly important for perennial grasses grown as primary crops with no market.

As of October 2010, there was no crop insurance available for perennial grasses, and this was thought to be a potential problem for new growers.[9] Crop insurance programs for perennial grasses will only be developed once a base of acreage is planted, but many producers may be reluctant to plant perennial grasses in the absence of crop insurance.[10] Switchgrass is currently being grown by the University of Tennessee Biofuels Initiative and by the Oklahoma Bioenergy Center. In Tennessee, 725 acres of switchgrass were planted during spring 2008 and an additional 1,900 acres were planted during spring 2009.[11] Producers accepted into the program are paid at a rate of $450 per acre of switchgrass per year for a three-year contract. An additional 3,000 acres were expected to be planted in 2010. In Oklahoma, 1,100 acres of switchgrass were planted in spring 2008, with 1,000 acres in a single tract of land, which is the largest area of switchgrass planting in the world. [12]

A stand of switchgrass will grow annually for 10 years, while a stand of *Miscanthus* may grow annually for 15 to 20 years.[13] Perennial grasses must first be established and may not be harvested during the first year. It may also take a few harvests before the grasses reach their peak yields. Deep roots help the plant store resources from year to year, which reduce necessary fertilizer inputs. Perennial grasses are also able to take advantage of long growing seasons. This is because planting and germination do not take place annually and the plants are above the ground and taking in sun for more days starting in the spring and going until early fall.

Due to their nutrient storage ability, these grasses can be grown on marginal cropland that is often considered unfit for growing corn or soybeans. However, simply because these grasses will grow on marginal land does not mean that they will yield well, and lower yields will ultimately increase the production cost per ton. Recent studies from the University of Tennessee arrived at contrary conclusions on this matter, with one finding that a biofuels processing plant would look for switchgrass from more productive soil types in order to minimize delivered costs, while the other found that producers would choose to plant switchgrass on less productive soils, since more productive soils are reserved for corn production. [14] This contrast reflects the differences in objectives from the perspectives of producers and biofuels processing plants.

Most early *Miscanthus* research has taken place in Europe, but some research plots are now being grown in the United States. *Miscanthus* is expected to yield well in the same locations where switchgrass yields well. However, field experiments in Illinois and Iowa conducted thus far indicate that switchgrass yields can be as little as one-fourth of *Miscanthus* yields.[15] *Miscanthus* yields are higher than switchgrass yields due to their larger mass, taller height, and longer growing season. Yields for both switchgrass and *Miscanthus* are high relative to agricultural residues. This higher yield serves to decrease the production cost per ton, but perennial grasses are still more costly to produce than residues, because they are dedicated crops that must be established and maintained. Current estimates for yields tend to be reported from small-scale research trials, where establishment may be more likely to succeed within the first year and allow for a first-year cutting. It is uncertain whether perennial grass yields will be as high once it is grown commercially by producers with limited switchgrass production experience.[16] Switchgrass is planted as a seed, while the most productive variety of *Miscanthus* does not produce seeds and must be planted as rhizomes, which makes establishment costs much higher. *Miscanthus* will also require more machine power to harvest because the quantity of biomass per acre is so much larger than with switchgrass. Despite establishment and harvesting costs being higher for *Miscanthus*, high yields allow those costs to be spread out over more tons of biomass per acre than switchgrass, which serves to lower the production cost per ton of *Miscanthus*. Further discussion of yields and production costs are in the sections that follow.

In the United States, perennial grasses could likely be grown on land that is currently not in use or is part of the Conservation Reserve Program (CRP), which protects land from erosion and environmental damage by maintaining vegetative cover such as native grasses.[17] For grasses, in particular, to be a viable feedstock for cellulosic biofuels, producers of these fairly uncommon crops will need to assume the risk of growing a long-run crop with high upfront establishment costs and possibly having to replant. Studies have found that switchgrass must be replanted 23% to 25% of the time due to seed dormancy and mistakes with establishment,[18] in which case establishment costs would increase substantially and overall per ton production costs would increase. Risk associated with high upfront establishment costs may be addressed through government grants to help with establishment costs, university planting programs that provide funding and expertise regarding planting and maintenance, or contracts with area cellulosic biofuels plants to supply biomass. The profitability of these relatively unknown crops in large-scale production, the profitability of other crops that could be grown in place of perennial grasses, and the proximity of production facilities will be important in determining the extent to which they are planted. Many producers who will enter into contracts with plants may be small and/or part-time operators who may lack the education, expertise, and equipment necessary to establish and maintain a perennial grass stand in a cost-effective manner. These producers may also be growing the grasses on small and dispersed fields, which will make management more expensive and less efficient. It is expected that land grant university extension programs will play a role in helping producers streamline and perfect the growing process once commercial production becomes more common. [19]

Agricultural Residues

Agricultural residues, such as corn stover, which includes the stalks, leaves, and cobs, are the byproducts left after harvest of crops that are already being planted. Residues are more readily available than perennial grasses, which need to be established, and are typically put forth as a less expensive feedstock option, because their establishment cost is attributed to the initial crop. Instead, the primary costs associated with residues are nutrient replacement, harvesting, storage, and transportation. For the purposes of this discussion, corn stover will be used as the primary residue.

Crop residues serve to prevent erosion and nutrient loss while a field is fallow, and excess removal of residue can compromise that protection and reduce soil organic matter. The appropriate amount to remove will depend on tillage practices, crop rotation, soil type, and topography. Extensive literature exists regarding the effect of residue removal on the health of the soil. Generally, more residue removal is thought to reduce organic matter and leave the soil susceptible to erosion. However, no one conclusion has been reached on the maximum amount that can be safely removed.[20] In fact, with increasing corn yields and residues roughly equal in weight to corn, some have argued that residue removal may be necessary as residue amounts increase. The amount of residue removed will depend, in large part, on the equipment used to remove the residue. For example, if conventional hay equipment is used, the amount of residue removed can vary depending on the number of passes. Baling alone will collect about 38% of residue, raking and baling will collect about 5 2.5% of residue, and shredding, raking, and baling could collect 70% of residue.[21]

Work remains to find the most efficient residue collection technique. For example, making multiple passes through the field—first to harvest the main crop, then to shred, rake, bale, and collect the residue—will increase the cost of collecting the residue and increase soil compaction in the field. In contrast, a single-pass system that performs several functions simultaneously may be overly slow to collect the principal crop and run the risk of unanticipated inclement harvest- time weather. More efficient collection technologies will be capital intensive and will likely be adopted by larger producers, which will leave smaller producers to use existing hay equipment or hire a custom operator to harvest residues.

Time is a critical factor for farmers in the fall harvest period. Harvest of residues must take place within a fairly small window after corn has been harvested and will be highly dependent on weather conditions. The top priority of producers will certainly be the corn crop, and if removing residue puts their corn yield at risk, producers may be reluctant to agree to remove it.

A one-pass harvest system that attaches to the combine appears to be the next step in equipment development for residue harvest if it can be done without slowing down the harvest. Hay equipment is available but less efficient, and equipment that will only collect residues may not appear for another 15 to 20 years.[22] Current research at Iowa State University is developing a one- pass harvesting system that attaches to a conventional combine as a modified header in the front and a chopper and blower in the back. This was preferred to a one-pass system that harvests the grain and the residue at the same time in a single stream, because additional equipment would be needed to separate the grain and the residue, and this mixture of crops may change a producer's eligibility for government programs and crop insurance. Stover is chopped into two inch pieces and blown into a wagon running alongside the combine. Currently, this attachment system is estimated to cost between $35,000 and $50,000.[23] The system is equipped with a switch to shut off the attachment and allow the residue to be left in a windrow behind the combine should residue harvest interfere with or slow grain harvest. Work continues to ensure that the use of this system does not slow down conventional grain harvest, regardless of the amount of residue the producer chooses to remove.

A one-pass harvesting system allows for residue to be harvested with the grain and keeps the residue from ever touching the ground, which may cause soil contamination and make conversion to liquid fuels less efficient.. However, a current shortcoming of this method is that residue coming directly from the combine only has a density of about 3 to 4 pounds per cubic foot, while the density would need to be between 12 and 14 pounds per cubic foot to transport it efficiently and fill a truck to its allowed weight limit.[24]

Research is also being conducted to determine at what height residue should be cut. Corn plants tend to have more moisture in the lower part of the plant, making this portion less efficient for conversion as moisture must be removed prior to conversion and also best left on the field to protect against soil erosion. The upper part and the cobs, however, have lower moisture contents and are most suitable for producing biofuels and making pretreatment and processing as efficient as possible.[25]

While residues are convenient because they accompany the primary crop, there is no reason that they must be removed and no long term commitment to do so unless the producer enters into a contract with the plant. Perennial grasses, however, are a commitment that must be harvested each year for the life of the stand.

Dedicated Tree Crops

Short-rotation woody crops grown for biomass on agricultural or other open land may include hardwood varieties such as poplar and willow. These trees are commercially grown as a crop and are adaptable to many different regions throughout the country. Short-rotation woody crops are attractive as a biomass feedstock due to quick regrowth following harvest. Some trees are already being harvested to make pulp and other small wood products. Short-rotation woody crops also provide environmental benefits such as low inputs, improved soil and water quality, reduced CO_2 emissions, and enhanced biodiversity.[26]

To establish short-rotation woody crops, cuttings from year old trees taken during the dormant season must be planted. In addition to high establishment costs (as with dedicated energy crops), it takes three to four years before the trees are ready to be harvested and begin yielding commercial returns. The life of the entire stand will be at least 20 years, and multiple (but not annual) harvests will take place. For example, seven to eight harvests may occur during the life of a willow tree crop, i.e., every three to four years.[27] During the growing time, though, very little annual maintenance is needed. Cost reduction for tree crop production will likely come from increased yields and production efficiency, as well as a mechanism for valuing environmental benefits.[28] Poplar and willow hybrids that will increase yield potential and reduce lignin content are currently being researched. Transportation efficiencies can also be achieved as stands of dedicated tree crops can be grown close to the conversion facilities and do not have to be trucked from commercial forests. Storage is less of an issue with tree crops because they can be stored on the stump and harvested as needed.

Forest Residues

Forest residues include naturally grown trees that may be of poor quality or too small to be used commercially, residues left in the forest after commercial logging, residues from clearing rotten trees that could cause forest fires, and residues from processing mills. Currently these residues, which are sustainable and plentiful, are burned, left in the forest to decay, or sent to landfills.[29] Using forest residues can serve to mitigate greenhouse gases, improve the health of forests, and avoid catastrophic fires and diseases. Collecting residues from within a forest, however, can be difficult, because efficient equipment has not been developed and most commercial logging operations are not set up to handle residues. Currently, the more efficient harvesting options chosen by large scale logging operations are methods that harvest round wood and biomass simultaneously. Similar to one-pass corn and corn stover harvesters, these systems do not require major changes to the current operation and do not add extra steps to the harvest process. However, one-step harvesters are capital intensive and are best suited for larger tracts of land.[30]

Forest biomass that does not come from a mill will be bulky, dirty, and high in moisture. Residues coming directly from the mill are advantageous, because they have lower moisture content and are in a more consistent form. However, these mill residues, which include bark, chunks of wood, shavings, and sawdust, are currently being used to create energy for processing mills or in other wood products, which may not leave much residue available for biofuel production.

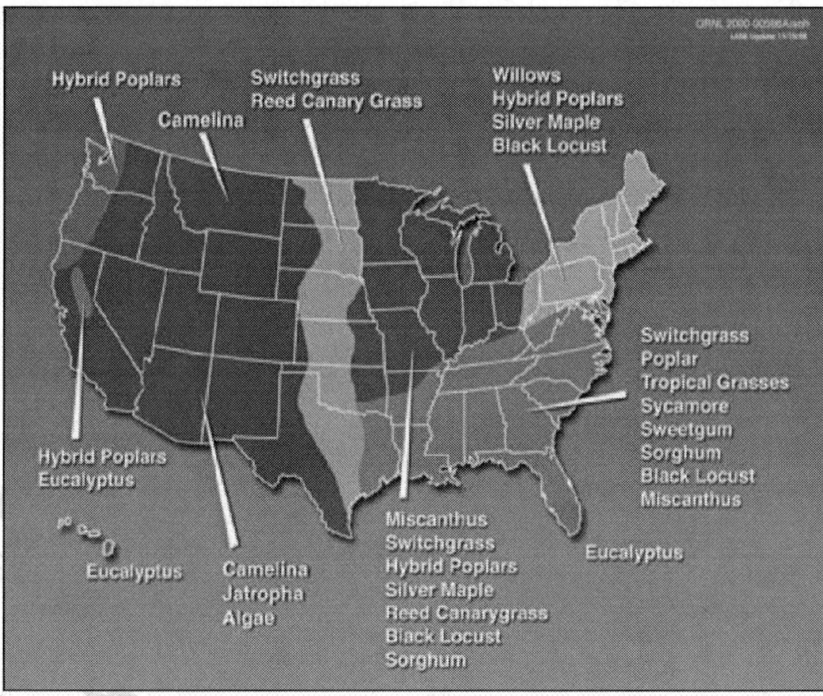

Source: U.S. Department of Energy, "Breaking the Biological Barriers to Cellulosic Ethanol: A Joint Research Agenda," A Research Roadmap Resulting from the Biomass to Bio fuels Workshop, Office of Energy Efficiency and Renewable Energy (Rockville, MD), 2006.

Figure 3. Expected Types of Biomass by Geographic Region in the US

The current language of the Energy Independence and Security Act of 2007 puts restrictions on the types of land residues can be collected for use as a cellulosic feedstock. Residues cannot be collected from federal lands, old-growth forests, or imperiled forests. It is uncertain whether the current limited definition would be a barrier to the production of cellulosic biofuels, as feedstocks vary in their distribution and ownership among regions. Woody biomass, for example, tends to be located on private land in the southeastern United States but on federal lands in the western United States.[31] However, H.R. 2454, the American Clean Energy and Security Act passed by the House of Representatives on June 26, 2009, would broaden the definition of renewable biomass to its Farm Bill definition, which allows renewable biomass to include that which is removed from federal lands.[32] The Senate has yet to consider the bill.

To deal with the bulky nature of forest biomass, it must be condensed by chipping, grinding, or bundling.[33] Chipping is the most efficient and least expensive method, but the knife blades can be damaged by dirt and other foreign material, which sometimes results in grinding being a preferred method. Forest biomass in both of these forms (chipped or ground) can be stored for several weeks but will eventually begin to decay. Bundling is the least efficient and most expensive option; however, bundles can be stored up to nine months with only a 10% loss.[34]

Potential Biomass Supply

In making plans to establish commercial scale cellulosic biofuels plants, knowledge of feedstock supply will be particularly important. However, with no substantial history of biomass crop production and residue collection, supply estimates cannot be based on past experience. Instead, some estimates of supply are based on what is possible and potential. It remains to be seen whether potential supply will accurately translate into actual supply. **Figure 3** shows what types of biomass are expected to come from different geographic regions in the United States. The eastern half of the country boasts the potential for more types of biomass, giving plants that locate there the option of using several types of feedstocks.

Figure 4 shows the distribution of biomass throughout the United States. Biomass is most densely located in the upper Midwest, Delta, Southeast, and Pacific Coast.

A 2005 study (the so-called "Billion-Ton Study") by the DOE and USDA found that through yield increase and the incorporation of perennial energy crops, forest and agricultural land in the United States could produce over 1.3 billion tons of biomass per year.[35] Of the 1.3 billion ton total in the 2005 study, 428 million dry tons would come from agricultural residues, 377 million dry tons would come from perennial crops, which include grasses and short rotation woody crops, and 368 million dry tons would come from forestlands. Of the 368 million dry tons from forestlands, 63 million dry tons come from logging residues, 147 million dry tons come from mill residues, and 59 million dry tons come from forest health removals.

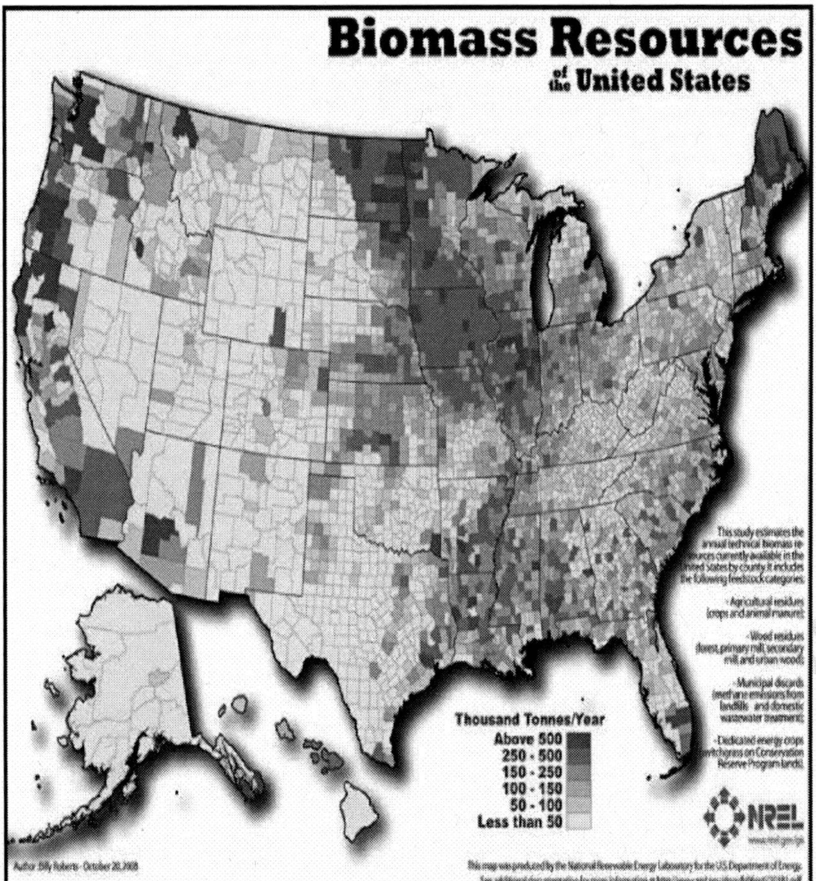

Source: Milbrandt, A., *A Geographic Perspective on the Current Biomass Resource Availability in the United States*, Dept. of Energy, National Renewable Energy Laboratory (Golden, CO), 2005.

Figure 4. Geographic Distribution of Biomass Resources in the United States

The rest comes from fuel wood harvests and urban wood waste. [36] The study generated considerable concern that its estimations were overly generous and optimistic, particularly as regards the availability of crop and forest residues and urban waste. As a result, the study is being updated, with the update scheduled for release by the end of 2010.

Figure 5 summarizes what portion of biofuels will be produced from each type of feedstock by the time the cellulosic biofuel industry has matured according to the "Billion-Ton Study." Crop residues, perennial grasses, and forest residues are expected to have nearly equal shares and to account for nearly all of biofuels production.

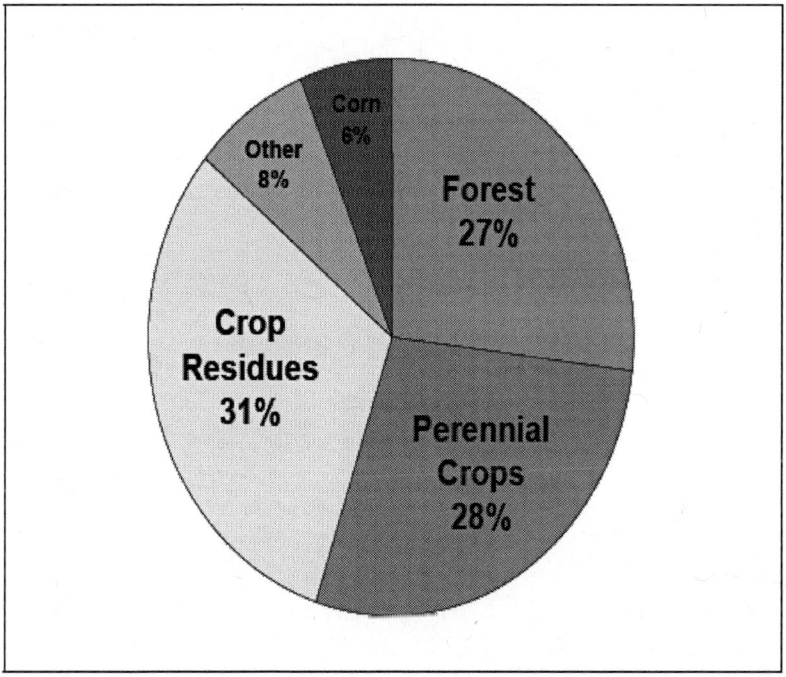

Source: Perlack et al, The *Billion-Ton Study* (2005).

Figure 5. Projected U.S. Biofuel Sources

Figure 5 summarizes what portion of biofuels will be produced from each type of feedstock by the time the cellulosic biofuel industry has matured according to the "Billion-Ton Study." Crop residues, perennial grasses, and forest residues are expected to have nearly equal shares and to account for nearly all of biofuels production.

In sharp contrast to the "Billion-Ton Study," the National Academy of Sciences found that with 2008 technologies and practices, a total of 416 million tons of biomass feedstocks could be harvested and produced sustainably for biofuel production, while 548 million tons will be available in 2020 due to more efficient use of land and increases in crop yields.[37] Of that 548 million tons, 366 million tons are expected to come from residues, 164 million tons are expected to come from dedicated energy crops, and 18 million tons are expected to come from yield increases in hay production.

The Environmental Protection Agency (EPA) has estimated that by 2022 agricultural residues will account for 5.7 billion gallons of cellulosic ethanol (4.9 billion gallons from corn stover), forestry biomass will account for 0.1 billion gallons, urban waste will account for 2.3 billion gallons, and dedicated energy crops will account for 7.9 billion gallons for a total of 16 billion gallons.[38] For corn stover, they assume that the ethanol conversion yield will be 92.3 gallons per dry ton, which equates to 53 million dry tons of corn stover being available. If the average harvested yield is 2 tons/acre (perhaps high), then 26.5 million acres of corn (or roughly about one-third of total annual harvested corn acres) would be harvested for stover.

A study based on 1997 data found that there were 36.2 million dry tons of logging residues available in the United States.[39] It was projected that this total would increase by 1.6% by 2010 and 5.1% by 2020. Additionally, a 2009 study from Purdue University found that cropland pasture and other idled land could provide up to 92.6 million acres of land on which to grow switchgrass and *Miscanthus* for cellulosic biofuels production.[40] This land could produce 327 million tons of switchgrass and 833 million tons of *Miscanthus* for a total of 17 billion to 53 billion gallons of potential ethanol production depending on the fraction of available land used.

These studies arrive at very different conclusions regarding the actual amount of available tons of biomass. BP, one of the major biofuel players in the oil industry, intends to use only dedicated crops. BP believes dedicated crops offer greater potential for achieving the scale of production they are seeking.[41] It will remain to be seen on what basis plants make their location decisions. These decisions could be driven by feedstock availability, plant construction, or fuel distribution, just to name a few potentially relevant considerations.

While supply estimates may indicate total availability of biomass, participation rates must also be considered, because the presence of biomass does not guarantee that landowners and producers will provide it to cellulosic biofuel plants. Studies have assumed participation rates ranging from 30% to 80%.[42] Participation rates will likely depend on the size of the plant, the price being paid for biomass by the plant, the cost of harvesting and collecting biomass for a producer of a given size, and the terms of contracts with producers. Participation rates may vary across areas. Differences in weather conditions that affect the ease of harvest also will be important.

Feedstock Production Yields

Feedstock yields have some degree of uncertainty as collection technologies and establishment and maintenance regimens develop. Yields will vary by geographic region, soil characteristics, and water availability. For corn stover, a distinction must be made between what is available and what is removable. Available stover will be a function of the corn yield while removable stover will be a function of available stover and the percentage of stover deemed appropriate for removal. A one-to-one ratio between corn stover and grain is usually assumed, which means there will be approximately 56 pounds of corn stover for every bushel of corn. Based on the grain yield, the available corn stover yield can be calculated. For example, assuming corn stover moisture content of 15%, grain yields of 125, 150, and 175 bushels per acre will result in 2.9, 3.5, and 4.1 dry tons of corn stover available per acre, respectively.[43] Corn stover moisture content can be estimated as a function of the grain moisture and the number of days after grain maturity.[44] After initial maturity, grain moisture may be around 40% and stover moisture may be as high as 75%. However, 80 days after maturity is reached grain moisture and stover moisture can both converge to around 10%. Once mature, the stover moisture will decline at a faster rate than the grain moisture. While the one-to-one stover to grain ratio is quite common, grain moisture above 18% may result in 0.8 to-one being a more realistic ratio for calculating the available corn stover yield.[45] Regardless of the stover to grain ratio, available corn stover will be a function of corn yield, and removable corn stover will be a function of the removal rate. Both corn yield and the appropriate removal rate will vary from field to field, which makes predicting the removable corn stover yield difficult. Overall, areas where grain yields are high will also have high residue yields, but the amount that is ultimately removed and used will depend on the ability of the producer to harvest or collect residues and how much can be removed while still maintaining the integrity and quality of the soil.

Table 2. Biomass Yields by Feedstock

Location	Yield	Source (Year)
	Corn Stover	
Midwest	3.6 available tons/acre	Sokhansanj, and Turhollow (2002)[a]
	1.7 removable tons/acre	
IN	4.25 available tons/acre	Brechbill and Tyner (2008)
	1.6 to 3.0 removable tons/acre (depending on harvesting technique)	
Midwest	3.6 available tons/acre	Sokhansanj, Turhollow, and Perlack (2002)[b]
	1.5 removable tons/acre	
Midwest	2.94 removable tons/acre	Quick (2003)[c]
IA	4 to 5 available tons/acre	Glassner et al (1998)[d]
	1.5 to 3.5 removable tons/acre	
IA and WI	1.25 to 1.55 removable tons/acre	Schechinger and Hettenhaus (1999)
IA	3.1 to 4.8 available tons/acre	Atchison, and Hettenhaus (2003)[e]
	2.2 to 3.3 removable tons/acre	
Not location specific	3.3 available tons/acre	Perlack and Turhollow (2002)
IA	1.1 removable tons/acre	Sokhansanj et al (2008)
	4.4 available tons/acre	
	2.9 removable tons/acre	
	Switchgrass	
IA	4.0 tons/acre	Duffy (2008)
OK	4.0 tons/acre	Epplin (1997)[f]
OK	3.75 to 6.50 tons/acre	Epplin et al (2007)[g]
TN	6.45 tons/acre	Garland (2008)
IL	2.4 tons/acre	Khanna (2008)
IL	4.2 tons/acre	Khanna et al (2008)[h]
ND, SD, NE	3.12 tons/acre	Perrin et al (2008)[i]
WI	4.0 to 5.8 tons/acre	Vadas et al (2008)[j]
	6.17 tons/acre (implied)	U.S. EPA (2009)[k]
IA	4.0 tons/acre	Duffy and Nanhou

		(2001)[l]
AR	5.0 tons/acre	Popp and Hogan (2007)[m]
IN	5.0 tons/acre	Brechbill and Tyner (2008)
MN, ND, SD	2.0 to 4.0 tons/acre	Tiffany et al (2006)[n]
ND	2.7 to 3.5 tons/acre	Bangsund et al (2008)[o]
IL	4.6 tons/acre	Heaton et al (2008)[p]
TN	4.3 to 8.8 tons/acre	Downing and Graham (1996)[q]
Southeast	7 to 16 tons/acre	Comis (2006)[r]
Western Corn Belt	5 to 6 tons/acre	Comis (2006)
ND	1 to 4 tons/acre	Comis (2006)
IA and IL	2.58 tons/acre	Khanna and Dhungana (2007)[s]
	Miscanthus	
IL	13 to 19 tons/acre	Khanna et al (2008)
IL	13.2 tons/acre	Heaton et al (2008)
IL	8.9 tons/acre	Khanna et al (2008)
Europe	4.5 to 13.4 tons/acre in central and northern Europe, up to 20 tons/acre in southern Europe	Lewandowski et al (2003)[t]
	Poplar/Willow	
TN	5 tons/acre/year	Mercker (2007)[u]
TN	2.4 to 4.3 tons/acre/year	Downing and Graham (1996)
MN	2.4 tons/acre/year	Downing (2004)[v]
MN	3.8 tons/acre/year	Lazarus (2008)[w]
MN	1.8 to 3.0 tons/acre/year	Updegraff, Baughman, and Taff (2004)[x]
IN	5 tons/acre/year	NAS (2009)[y]
NY	5 tons/acre/year	State Univ. of New York (undated)
	Hardwood Residues	
2009	20-40% of stem wood volume from conventional logging	Pelkki (2009)

Source: Compiled by Tyner, Brechbill, and Perkis, Purdue University, August 2010, from listed sources.

[a] Sokhansanj, S. and A. Turhollow, "Baseline Cost for Corn Stover Collection," *Applied Engineering in Agriculture*, 18(5), 2002, pp. 525-530, [hereafter referred to as Sokhansanj and Turhollow (2002)].

[b] Sokhansanj, S., A. Turhollow, and R. Perlack, "Stochastic Modeling of Costs of Corn Stover Costs Delivered to an Intermediate Storage Facility," *American Society of Agricultural Engineers Annual International Meeting*, CIGR XVth World Congress (Chicago, IL), 2002.

[c] Quick, G.R., "Single-Pass Corn and Stover Harvesters: Development and Performance," *Proceedings of the International Conference on Crop Harvesting and Processing*, ASABE Publication Number 701 P11 03e, American Society of Agricultural and Biological Engineers (Louisville, KY), 2003.

[d] Glassner, D.A., J.R. Hettenhaus, and T.M. Schechinger, "Corn Stover Collection Project," *BioEnergy '98: Expanding BioEnergy Partnerships*, 1998, pp. 1100-1110, [hereafter referred to as Glassner et al (1998)].

[e] Atchison, J.E. and J.R. Hettenhaus, *Innovative Methods for Corn Stover Collecting, Handling, Storing and Transporting*, NREL/SR-5 10-33893, National Renewable Energy Laboratory (Golden, CO), March 2003.

[f] Epplin, F.M., "Cost to produce and deliver switchgrass biomass to an ethanol-conversion facility in the southern plains of the United States," *Biomass and Bioenergy*, 11(6), 1997, pp. 459-467.

[g] Epplin, F.M., C.D. Clark, R.K. Roberts, and S. Hwang , "Challenges to the Development of a Dedicated Energy Crop," *American Jl of Agr. Economics* 85(5), 2007, pp. 1296-1302, [hereafter referred to as Epplin et al (2007)].

[h] Khanna, M., B. Dhungana, and J. Clifton-Brown, "Cost of producing miscanthus and switchgrass for bioenergy in Illinois," *Biomass and Bioenergy*, 32(6), 2008, pp. 482-493, [hereafter referred to as Khanna et al (2008)].

[i] Perrin, R., K. Vogel, M. Schmer, and R. Mitchell., "Farm-Scale Production Cost of Switchgrass for Biomass," *Bioenergy Research* Vol. 1, No. 1, 2008, pp. 91-97.

[j] Vadas, P.A., K.H. Barnett, and D.J. Undersander, "Economics and Energy of Ethanol Production from Alfalfa, Corn, and Switchgrass in the Upper Midwest, USA," *Bioenergy Research* Vol. 1, No. 1, 2008, pp. 44-55.

[k] U.S. EPA, *Draft Regulatory Impact Analysis: Changes to the Renewable Fuel Standard Program*, EPA-420-D-09- 001, Office of Transportation and Air Quality, May 2009, [hereafter referred to as U.S. EPA (2009)].

[l] Duffy, M. and V. Nanhou, *Costs of Producing Switchgrass for Biomass in Southern Iowa*, PM 1866, University Extension, Iowa State University (Ames, IA), April 2001.

[m] Popp, M. and J. R. Hogan, "Assessment of Two Alternative Switchgrass Harvest Transport Methods," *Biofuels, Food, and Feed Tradeoffs*, Farm Foundation Conference (St. Louis, MO), 2007.

[n] Tiffany, D.G., B. Jordan, E. Dietrich, B. Vargo-Daggett., *Energy and Chemicals from Native Grasses: Production, Transportation and Processing Technologies Considered in the Northern Great Plains*, Staff Paper P06-11, University of

Minnesota, Department of Applied Economics, College of Food, Agricultural & Natural Resource Sciences, 2006.

[o] Bangsund, D.A., E.A. DeVuyst, and F.L. Leistritz, "Evaluation of Breakeven Farm-gate Switchgrass Prices in South Central North Dakota," *Agribusiness and Applied Econ.s* Report No. 632-S, N. Dak. St. Univ. (Fargo, ND), Aug. 2008.

[p] Heaton, E.A., F.G. Dohleman, and S.P. Long, "Meeting US biofuel goals with less land: the potential of Miscanthus," *Global Change Biology* 14(9), Sept. 2008, pp. 2000-20 14, [hereafter referred to as Heaton et al (2008)].

[q] Downing, M. and R.L. Graham, "The Potential Supply and Cost of Biomass From Energy Crops in the Tennessee Valley Authority Region," *Biomass and Bioenergy* 11(4), 1996, pp. 283-303, [hereaft. Downing and Graham (1996)].

[r] Comis, D., "Switching to Switchgrass makes Sense," *Agricultural Research*, Agricultural Research Service, USDA, 2006; [hereafter referred to as Comis (2006)].

[s] Khanna, M. and B. Dhungana., "Economics of alternative feedstocks," Chapter 8, *Corn-based ethanol in Illinois and the US: A Report from Department of Agricultural and Consumer Economics*, University of Illinois, University of Illinois (Urbana-Champaign, IL), 2007, pp. 129-146.

[t] Lewandowski, I., J. Scurlock, E. Lindvall, and M. Christou, "The Development and Current Status of Perennial Rhizomatous Grasses as Energy Crops in the US and Europe," *Biomass and Bioenergy* 25(4), 2003, pp. 335-361.

[u] Mercker, D., *Short Rotation Woody Crops for Biofuel*, SP702-C, University of Tennessee Extension, University of Tennessee Biofuels Initiative, University of Tennessee (Knoxville, TN), 2007.

[v] Downing, M., *Hybrid Poplar Production in Minnesota on a Large Scale, in Agriculture as a Producer and Consumer of Energy*, Farm Foundation Conference (Arlington, Virginia), 2004.

[w] Lazarus, W., *Energy Crop Production Costs and Breakeven Prices Under Minnesota Conditions*, Staff Paper P08-11, Department of Applied Economics, Univ. of Minnesota, 2008, [hereafter referred to as Lazarus (2008)].

[x] Updegraff, K., M.J. Baughman, and S.J. Taff, "Environmental benefits of cropland conversion to hybrid poplar: economic and policy considerations," *Biomass and Bioenergy* 27(5), 2004, pp. 411-428.

[y] NAS, *Liquid Transportation Fuels from Coal to Biomass* (2009).

Perennial grasses will have lower yields in the years immediately after establishment and will then increase to peak yield once mature. There is also the chance that seeds may remain dormant and do not grow after planting. This will require the grasses to be planted again, which doubles the establishment costs and delays the first harvest and the eventual peak yield. Perennial grass yields will tend to be higher in regions where temperatures are high and winters are short in order to provide a longer harvest window and longer growing season.[46]

Forest residue yields are a function of yield from conventional logging. For hardwood stands, 20% to 40% of the initial yield can be recovered as additional residue.[47] As with corn stover, collection technology and the available resources of a given type and size of operation will impact the amount of residue collected.

Table 2 summarizes the estimates of biomass yield from several studies in different geographic regions. A majority of these yields are from research test plots, small-scale producer experiments, or based on assumptions for the region. As with yields for other crops, biomass yields will not be uniform and constant across a given area. Much variation will depend on weather conditions, soil types, topography, and producer expertise and experience. For corn stover, up to 5 tons per acre could be available but not removable. Switchgrass yields are between 3 and 6 tons per acre, with yields going above that in warmer climates. *Miscanthus* seems to average about 13 tons per acre on the few sites it has been planted but will increase in warmer climates as well. Short-rotation woody crops yield 2 to 5 tons per acre per year. Forest residue yields are calculated as a percentage of the yield of the forest stand.

Production Costs

Several studies estimate the production costs of different types of biomass. These cost estimates can vary for a wide variety of reasons. The location and timing of the study can influence the cost of inputs, labor, and equipment. Yields will affect the number of tons over which to spread per acre costs. Many of the major differences among cost studies relate to assumptions regarding what to exclude and include in the calculations. Explicitly identifying the assumptions and parameters of differing studies can often explain discrepancies. No one set of assumptions and parameters is thought to be universally correct.

Corn stover production costs primarily include collection after corn grain harvest and nutrient replacement. What tends to make corn stover a less expensive biomass feedstock is that it is a byproduct. All establishment costs are accounted for in corn grain production. When left on the ground, corn stover provides protection from soil erosion and serves to retain nutrients while land is fallow. Depending on the amount of stover removed after harvest, additional nutrients will need to be added before planting the following year. Fertilizer costs will increase with energy costs, which will also increase the cost of corn stover. In addition, with increasing corn yields, it may be useful to remove part of the stover to prepare the field for crop operations the next year.

Table 3 breaks down the production costs for corn stover from several studies. The farm gate cost per ton ranges from $12 to $67. Low estimates are often from older studies that assumed lower fertilizer and energy costs. Some studies also did not consider nutrient replacement costs at all, which leads to lower total costs. The total cost in some instances only considers harvest and nutrient replacement, while other studies include a payment to the producer and assume some storage loss.

Table 3. Corn Stover Production Costs

Source (Year)	Location	Harvest ($/ton)	Fertility Replacement ($/ton)	Payment to Land Owner or Farmer ($/ton)	Harvestable Yield (tons/acre)	Farm Gate Cost ($/ton)
Gallagher et al (2003)a	KS	$5.96	$6.47	N/A	3.33	$12
Gallagher et al (2003)	IA	$6.27	$6.46	N/A	3.13	$13
Glassner et al (1998)	IA	$14.60	N/A	$3-$15	1.5-3.0	$18-30
Sokhansanj and	Midwest	$20-$22	N/A	N/A	1.7	$20-$22
Turhollow (2002)						
Sokhansanj et al (2008)	IA	$21.95	N/A	N/A	2.9	$24
Graham et al (2007)[b]	US	$18-$33	$6.50	N/A	1.4-2.3	$25-$40
Brechbill and Tyner (2008)	IN	$5.88	$15.64	15% of per ton cost	1.6-3.0	$35
Perlack and Turhollow (2002)	Not location specific	$22.30	Covered by payment to farmer	$10	1.1	$35-$37
EPA (2009)	IN	$23.73	$11.81	$10	2.0	$43-$46

Table 3. (Continued)

Petrolia (2008) and Eidman et al (2009)[c]	MN	$20	$4.21 for 2000-2004 prices	$20	1.25-1.55	$53-$56
			$10.64 for 2007 prices			
Aden et al (2002)[d]	IA	$26	$7	$10	2.2	$56
Khanna (2008)	IL	$35.05	$8.27	$24	1.85	$67

Source: Compiled by Tyner, Brechbill, and Perkis, Purdue University, August 2010, from listed sources.

[a] Gallagher, P., M. Dikeman, J. Fritz, E. Wailes, W. Gauther, and H. Shapouri., *Biomass from Crop Residues: Cost and Supply Estimates, Agricultural Economic* Report No. 819, U.S. Department of Agriculture, Office of the Chief Economist, Office of Energy Policy and New Uses, February 2003, [hereafter referred to as Gallagher et al (2003)].

[b]Graham, R.L., R. Nelson, J. Sheehan, R. D. Perlack, and L. L. Wright., "Current and Potential US Corn Stover Supplies," *Agronomy Journal* 99(1), 2007, pp. 1 - 11.

[c] Eidman, V., D. Petrolia, H. Huang, and S. Ramaswamy. *The Economic Feasibility of Producing Ethanol from Corn Stover and Hardwood in Minnesota*, Staff Paper P09-3, Department of Applied Economics, University of Minnesota, 2009, [hereafter referred to as Eidman et al (2009)].

[d] Aden, A., M. Ruth, K. Ibsen, J. Jechura, K. Neeves, J. Sheehan, and B. Wallace., L. Montague, A. Slayton, and J. Lukas, *Lignocellulosic Biomass to Ethanol Process Design and Economics Utilizing Co-Current Dilute Acid Prehydrolysis and Enzymatic Hydrolysis for Corn Stover*, NREL/TP-5 10-32438, National Renewable Energy Laboratory (Golden, CO), U.S. Department of Energy, June 2002.

Table 4 outlines the method and some assumptions used in two recent corn stover cost estimate studies that arrive at very different estimates for the per ton cost of corn stover. A few differences to note include yield, nutrient replacement, storage, densification, and payment to the producer.

- Because the Brechbill and Tyner study allows for higher removal rates, the yield is higher, which helps decrease per ton costs.
- The Eidman and Petrolia studies do not include nitrogen costs in nutrient replacement, because they assume a corn-soybean rotation.
- The Brechbill and Tyner study assume bales of stover will be stored along the edge of the field until needed by the plant, while the Eidman

and Petrolia studies assume that bales are transported to a regional storage facility and sometimes even stored under roof. The additional transportation and the construction of a storage facility serve to increase per ton costs.

- The Eidman and Petrolia studies also allow for the possibility of densification just before the stover is transported to the plant. This serves to increase the density of the stover and reduce transportation costs by allowing more to be loaded onto each truck. The Brechbill and Tyner study did not consider this and assumed that stover would be hauled to the plant in bale form.
- Finally, Brechbill and Tyner provide a payment to the producer that is 15% of the cost of production. In the case of corn stover, this is approximately $5 per dry ton. The Eidman and Petrolia studies pay $20 per dry ton.

Table 4. Assumptions and Parameters Used in Two Corn Stover Cost Studies

	Sources	
Item	**Brechbill and Tyner (2008)**	**Petrolia (2008); Petrolia (2009), and Eidman et al (2009)**
Location	Indiana	Minnesota, Iowa, South Dakota
Yield	4.25 available tons per acre with 1.6 to 3.0 removable tons per acre	One-to-one stover to grain ratio, grain yields vary depending on the specific site, approximately 1.3 to 1.6 removable tons per acre
Nutrient replacement	Replace nitrogen, phosphorus, and potassium at a cost of $15.64 per ton of stover removed	Replace phosphorus and potassium at a cost of $4.21 per ton of stover removed based on 2000 to 2004 average fertilizer prices and $10.64 per ton of stover removed based on 2007 fertilizer prices
Participation rate	50% and 75%	50%
Harvest method	Baling only, raking and baling, or shredding, raking, and baling	Baling only and shredding, raking, and baling
Equipment	Owned equipment for farm sizes of 500, 1,000, 1,500, and 2,000 acres or custom hired operators	Purchased shredder, rake, baler, bale picker, and telehandler
Removal rate	38% for baling only, 52.5% for raking and baling, and 70% for shredding, raking, and baling	30% when using round bales, 40% when using square bales

Table 4. (Continued)

Baling	1,000 pound round bales wrapped in either twine, net wrap, or plastic wrap	880 pound round bales wrapped in plastic mesh and 1,598 pound rectangular bales wrapped in twine
Storage	Bales stored at the edge of the field until needed by the plant, storage premium paid after six months of storage	Round bales stored outdoors since wrapped with plastic, rectangular bales stored indoors
Dry matter loss	Depends on method used to bale the stover, 3.13% per month of storage when using twine, 1.4% per month of storage when using net wrap, 1.025% per month of storage when using plastic wrap	2% storage loss for both types of bales
Densification	Not applicable	Densification facility is located next to the central storage facility and densification would take place immediately before stover is taken to the plant. Results in significant
		cost reduction in transportation of round bales and only slight cost reduction in transportation of rectangular bales.
Transportation	Flatbed semi-trucks, owned equipment for farm sizes of 500, 1,000, 1,500, and 2,000 acres or custom hired operators, bales transported directly from the edge of the field to the plant when requested by the plant	Bales transported via flatbed semi-truck from the edge of the field to a central storage facility
Payment to producer	15% of the per ton product cost	$20 per dry ton
Farm gate cost (not including transport)	$35 per dry ton	Approximately $53 per dry ton for rectangular bales, $56 per dry ton for round bales, and $76 per dry ton when densified
Transportation	13 dry tons per loaded trailer $2.60 per loaded mile $0.20 per mile per dry ton For a 13 dry ton load travelling 25 miles, $5.00 per dry ton in total transportation cost	22.4 dry tons per loaded trailer with rectangular bales 11.9 dry tons per loaded trailer with round bales 23 dry tons per loaded trailer with densification $2.82 per loaded mile for 0 to 25 miles $2.22 per loaded mile for 26 to 100 miles $1.96 per loaded mile for over 100 miles

		For rectangular bales travelling 25 miles, $3.15 per dry tons in total transportation cost For round bales travelling 25 miles, $5.92 per dry tons in total transportation cost For densified bales travelling 25 miles, $3.07 per dry tons in total transp. Cost
Total delivered cost (25 miles)	$40 per dry ton	$56.15 per dry ton (rectangular bales) $61.92 per dry ton (round bales) $79.07 per dry ton (densified)

Source: Compiled by Tyner, Brechbill, and Perkis, Purdue University, August 2010, from listed sources.

Even without seeing the exact calculations involved in each of these studies, breaking down the assumptions of the studies can help determine the source of differences in total costs. This same exercise can be done with switchgrass studies, and differences in land rent, establishment, maintenance, harvest, and storage can account for major total cost differences.

Perennial grasses production costs are comprised of both one-time establishment and annual maintenance costs. Yields for these grasses may take up to three years to reach their peak, which results in high costs and low benefits initially. Growing perennial grasses is also a relatively long term commitment of resources. Deciding to plant perennial grasses means harvesting each year for the life of the stand.

Table 5. Switchgrass Production Costs

Source (Year)	Location	Land Cost ($/acre)	Harvest Method	Harvestable Yield (tons/acre)	Farm-Gate Cost ($/ton)
Epplin (1997)	OK	$30	Large round bales	4.0	$23
Downing and Graham (1996)	TN	Various	Not specified	4.3-8.8	$28 to $64
Epplin et al (2007)	OK	$60	Large rectangular bales	3.75-6.5	$37-$53

Table 5. (Continued)

Bangsund et al (2008)	ND	$0	Not specified	2.7-3.5	$47 to $76
Khanna et al (2008)	IL	$0	Large rectangular bales	4.2	$52
Mooney et al (2008)[a]	TN	$100	Large round bales	8.83	$53
Brechbill and Tyner (2008)	IL	$70	Large round bales	5.0	$55
Garland (2008)	TN	$0	Large round bales	6.45	$62
Ferland (2001)[b]	GA	$20	Not specified	6.0	$66
Carpenter and Mees (2008)[c]	MO	$33	Not specified	4.5	$86
Khanna (2008) based on calculations by Epplin (2009)	IL	$77	Not specified	2.4	$113 (includes foregone profits from a corn and soybean rotation)
Duffy (2008)	IA	$80	Large square bales	4.0	$114 (includes transportation to storage, storage in a building, and transportation to the plant)

Source: Compiled by Tyner, Brechbill, and Perkis, Purdue University, August 2010, from listed sources.

a. Mooney, D.F., et al., “Switchgrass Production in Marginal Environments: A Comparative Economic Analysis across Four West Tennessee Landscapes,” selected paper, *American Agricultural Economics Association Annual Meeting*, Orlando, FL, 2008.

b. Ferland, C., Switchgrass for Co-genernation Fuel Feasibility, University of Georgia, Center for Agribusiness and Economic Development, Dept. of Agricultural and Applied Economics, University of Georgia (Athens, GA), 2001.

c. Carpenter, B. and M. Mees, “Bale, Silage, and Bio-energy, Projected Budgets for 2009,” *Agricultural Electronic Bulletin Board, University of Missouri Extension*, 2008.

Table 5 breaks down the production costs for switchgrass from several studies. The farm gate cost per ton ranges from $23 to $114. A major factor in the difference among the estimates is the assumed yield, which for the studies mentioned below ranges from 2.4 to 8.8 tons per acre. Another factor accounting for differences in switchgrass production costs is assumptions made regarding land rent.

Table 6 lists the production cost estimates for *Miscanthus* from a study conducted at the University of Illinois, Presently, there is not a significant amount of production cost estimates for *Miscanthus*, but many of the same points hold as did with switchgrass. This study finds that *Miscanthus* will cost $38 per ton, which is closer to estimates for corn stover than for switchgrass. The major difference relative to switchgrass is the substantially higher yields, which leads to a lower per ton cost for *Miscanthus.*

Table 6. Miscanthus Production Costs

Source (Year)	Location	Land Cost ($/acre)	Harvest Method	Harvestable Yield (tons/acre)	Farm-Gate Cost ($/ton)
Khanna et al (2008)	IL	$0	Large rectangular bales	13-19	$38

Source: Compiled by Tyner, Brechbill, and Perkis, Purdue University, August 2010, from listed sources.

Table 7 summarizes the production costs for short-rotation woody crops. Total costs will depend on what yield is achieved and the type of land on which the trees are established.

Table 7. Short-Rotation Woody Crop Production Costs

Source (Year)	Location	Land Cost ($/acre)	Harvest Method	Harvestable Yield (tons/acre/yr)	Farm-Gate Cost ($/ton)
Downing and Graham (1996)	TN	Various	$17	2.4 to 4.3	$29-$46 on former cropland, $44-$63 on former pasture
State Univ. of New York (2008)	NY	$35	$15.10	5	$52 for 13 yr rotation, $47 for 22 yr rotation
Han (2009)	IN	$90	$15.05	5	$57
Lazarus (2008)	MN	$40	Not specified	3.8	$63

Source: Compiled by Tyner, Brechbill, and Perkis, Purdue University, August 2010, from listed sources.

Table 7 summarizes the production costs for short-rotation woody crops. Total costs will depend on what yield is achieved and the type of land on which the trees are established.

Short rotation woody crops are similar to perennial grasses in the way that they are produced. A majority of the cost is incurred in site preparation, establishment, and in those years when harvest takes place. Harvest, however, does not take place annually.

Table 8. Forest Residue Production Rates and Costs

Technology	Production Rates	Cost per Green Ton	Approximate Cost per Dry Ton
Chipping	300-400 tons/day	$8-$12	$11.60-$17.40
Grinding	250-325 tons/day	$10-$12	$14.50-$17.40
Bundling/Baling	100-200 tons/day	$12-$20	$17.40-$29.00

Source: Pelkki (2009).

Notes: These costs are cited in terms of green tons. A dry ton will have a moisture content of less than 10%, while a green ton will have a moisture content of 40 to 50% just after harvest. Dry ton costs shown above assume 45% moisture content.

Table 8 outlines costs associated with processing forest residues that come directly from the forest. This only includes harvesting to make the forest residues more uniform and does not include transportation costs.

Capital investments for systems to harvest and process forest residues are estimated to cost between $800,000 and $1,200,000.[48] Another study from Minnesota found the delivered cost of hardwood residue to be $40.37 per dry ton plus the cost of transportation.[49]

Table 9. Conventional Non-Uniform Feedstock Production Costs ($/dry ton)

Biomass Type	Harvest and Collection	Storage	Handling and Transport	Receiving and Preprocessing	Average Total Cost
Corn Stover	$21.61 ± $2.69	$8.11 ± $0.66	$11.93 ± $1.25	$13.74 ± $1.31	$55.39
Switchgrass	$14.92 ± $1.45	$7.08 ± $0.52	$14.13 ± $1.43	$13.47 ± $1.30	$49.60

Source: Hess et al (2009).

Note: Data presented as (Mean Value) +/- (Standard Deviation) were derived from Monte Carlo Simulation of a stochastic model. The average cost for a biomass type is the sum of mean values across the four categories.

Two Alternate Cost Studies from 2009

The Idaho National Laboratory and the National Academy of Sciences released studies in 2009 that provide a different perspective and more comprehensive assessment of feedstock costs. This section provides a summary and assessment of these studies.

Idaho National Laboratory Monte Carlo Simulation Results

The Idaho National Laboratory[50] looks at three feedstock supply systems; one system produces a non-uniform feedstock[51] and the other two systems produce uniform feedstocks.[52] In order to compare to the other cost estimates already cited, **Table 9** outlines the estimates for the major components of a conventional production process with a non-uniform feedstock for both corn stover and switchgrass. This production technology is meant to represent technology that is available today and suggests that switchgrass costs about $50 per dry ton compared with about $55 per dry ton for corn stover.

National Academy of Sciences Feedstock Costs

The National Academy of Sciences report[53] has estimated biofuel feedstock costs based to some degree on parameters used in the literature and to some degree on expectations about demand for biomass in a mature industry. These estimates are intended to represent the willingness-to-accept price for the last delivered dry ton of biomass and are assumed to be equivalent to the marginal cost of production for the last ton. With an upward sloping marginal cost curve, each additional ton of biomass costs more than the previous ton due to increasing transportation costs. A plant will likely contract its supply and cover the transportation costs, and each producer will receive the farm-gate price, regardless of their distance from the plant. It is assumed that these prices will apply to a mature industry that demands 500 million tons of biomass per year. Assuming a yield of 70 gallons per ton, this will equate to a 35 billion gallon cellulosic biofuel industry. This suggests that plants will have larger capacities and demand biomass from much further distances than currently planned plants or those that will first emerge on a commercial scale.

The National Academy of Sciences panel expects that feedstock costs will decline over time with improvements in crop yields, land management, and logistics, such as handling, storage, and transportation. Cost estimates were done for corn stover, switchgrass, *Miscanthus,* prairie grasses, woody biomass, and wheat straw. Overall, the total feedstock costs determined by the National

Academy of Sciences study are higher than most of those predicted in the literature. The following discussion is an attempt at breaking down these costs to determine how they were calculated and what portions of the total costs are particularly higher than expected.

Estimates for corn stover, switchgrass, *Miscanthus*, and woody biomass are presented in the following four tables (**Table 10**, **Table 11**, **Table 12**, and **Table 13**). Where applicable, cost estimates include establishment and seeding, nutrient replacement, harvesting and maintenance, storage, chipping, transportation, stumpage fees, and land and opportunity costs.

The results are presented for six estimated ranges: an estimate based on the current literature followed by a low, 50% low (i.e., average of "low" and "baseline"), baseline, 50% high (i.e., average of "high" and "baseline"), and high estimates. The baseline estimate is treated as the midpoint of the range of estimates, whereas the low and high cost scenarios are considered the best case and worse case scenarios, respectively. The 50% low and 50% high estimates are thought to be a more reasonable range. Baseline estimates are considered to be cost estimates for 2008, while the 50% low estimates are considered to apply to 2020.

Corn stover costs (**Table 10**) can be compared to those in **Table 3**. The highest nutrient replacement value found in the literature was around $15 per ton, but it is the baseline value in the National Academy study. The highest harvest cost found in the literature was around $35 per ton, but this is below the low cost value in this study. Compared to harvest and maintenance costs for switchgrass and *Miscanthus*, corn stover costs are assumed to be slightly higher due to requiring additional labor at the same time as grain harvest, which is likely an overstatement since most stover would be harvested after grain. In addition, land rent is included for corn stover as the opportunity cost of land, but since corn stover is a byproduct of corn, the land rent should not be applied to corn stover.

Switchgrass (**Table 11**) and *Miscanthus* (**Table 12**) costs can be compared to those in **Table 5** and **Table 6**, respectively. The land rent/opportunity cost is assumed to be between $150 and $250 per acre. While some perennial grasses may be planted on land that might also be used for corn or soybeans, it is expected that most grasses will initially be planted on more marginal land or pasture land, which likely would have substantially lower land rents. As a result, these land rent costs seem unreasonably high. Storage costs are mostly within the range estimated in the literature. Transportation distances may be slightly higher since assumptions regarding demand are for a mature industry.

Table 10. Willingness-to-Accept Corn Stover Price per Ton

	Literature Estimate	Low	50% Low	Baseline	50% High	High
Yield	2 to 3 tons/acre	2.5 tons/acre	2.25 tons/acre	2 tons/acre	1.75 tons/acre	1.5 tons/acre
Nutrient Replacement	$4 to $21/ton	$10/ton	$12.50/ ton	$15/ton	$17.50/ ton	$20/ton
Harvesting and Maintenance	Up to $35/ton	$40/ton	$42.50/ ton	$45/ton	$47.50/ ton	$50/ton
Storage	$2 to $17/ton	$10/ton	$12.50/ ton	$15/ton	$17.50/ ton	$20/ton
Transportation	$0.09 to $0.63/tn/mi or 22 to 67 miles	$0.25/tn/mi for 20 miles = $5/ton	$0.30/tn/mi for 25 miles= $7.50/ton	$0.35/tn/mi for 30 miles=$0.50/ton	$0.40/tn/mi for 35 miles= $14/ton	$0.45/tn/mi for 40 miles= $18/ton
Cropland Rental Rates	$0 to $143 per acre	$0/ac = $0/ton	$25/ac = $11/ton	$50/ac = $25/ton	$75/ac = $43/ton	$100/ac = $67/ton
Total Cost ($/ton)		**$65**	**$86**	**$110**	**$140**	**$175**

Source: "Biomass Resources for Liquid Transportation Fuels," Chapter 2, NAS, *Liquid Transportation Fuels from Coal to Biomass* (2009).

Table 11. Willingness-to-Accept Switchgrass Price per Ton

	Literature Estimate	Low	50% Low	Baseline	50% High	High
Yield	0.89 to 9.8 ton/acre	6 ton/acre	5 ton/acre	4 ton/acre	3 ton/acre	2 ton/acre
Establishment and Seeding	$30 to $200/acre converted to $/ton	$75/acre = $12.50/ton	$87.50/ac = $17.50/ton	$100/ac = $25/ton	$112.50/ac = $37.50/ton	$125/ac = $62.50/ton
Nutrient Replacement	$4 to $21/ton	$5/ton	$7.50/ton	$10/ton	$12.50/ton	$15/ton
Harvesting and Maintenance	na	$35/ton	$37.50/ton	$40/ton	$42.50/ton	$45/ton
Storage	$2 to $17/ton	$10/ton	$12.50/ton	$15/ton	$17.50/ton	$20/ton
Transportation	$0.09 to $0.63/ tn/mi for 22 to 67 miles	$0.25/tn/mi for 20 miles = $5/ton	$0.30/tn/mi for 25 miles=$7.50/ton	$0.35/tn/mi for 30 miles=$10.50/ton	$0.40/tn/mi for 35 miles=$14/ton	$0.45/tn/mi for 40 miles=$18/ton
Cropland Rental Rates	$76 to $230/acre	$150/ac = $25/ton	$175/ac = $35/ton	$200/ac = $50/ton	$225/ac =$75/ton	$250/ac = $125/tn
Total Cost ($/ton)		**$93**	**$118**	**$151**	**$199**	**$286**

Source: "Biomass Resources for Liquid Transportation Fuels," Chapter 2, NAS, *Liquid Transportation Fuels from Coal to Biomass* (2009).

Notes: na = not available.

Table 12. Willingness-to-Accept Miscanthus Price per Ton

	Literature Estimate	Low	50% Low	Baseline	50% High	High
Yield (tons/ acre)	3.4 to 17.8 tons per ac	12 ton/acre	10.5 ton/acre	9 ton/acre	7.5 ton/acre	6 ton/acre
Establishment and Seeding	$43 to $350 per ac	$175/ac=$14.58/ton	$200/ac= $19.05/ton	$225/ac = $25/ton	$250/ac=$33.33/ton	$275/ac=$45.83/ton
Nutrient Replacement	$4 to $21/ton	$5/ton	$7.50/ton	$10/ton	$12.50/ton	$15/ton
Harvesting and Maintenance	na	$35/ton	$37.50/ton	$40/ton	$42.50/ton	$45/ton
Storage	$2 to $17/ton	$10/ton	$12.50/ton	$15/ton	$17.50/ton	$20/ton
Transportation	$0.09 to $0.63/ton/ mile for 22 to 67 miles	$0.25/ton/mile for 20 miles = $5/ton	$0.30/ton/mile for 25 miles = $7.50/ton	$0.35/ton/mile for 30 miles=$10.50/ ton	$0.40/ton/mile for 35 miles = $14/ton	$0.45/ton/mile for 40 miles = $18/ton
Cropland Rental Rates	$76 to $230 per acre	$150/ac= $25/ton	$175/ac = $35/ton	$200/ac = $50/ton	$225/ac = $75/ton	$250/ac =$125/ton
Total Cost	**($/ton)**	**$82**	**$101**	**$123**	**$150**	**$186**

Source: "Biomass Resources for Liquid Transportation Fuels," Chapter 2, NAS, *Liquid Transportation Fuels from Coal to Biomass* (2009).

Table 13. Willingness-to-Accept Woody Biomass Price per Ton

	Literature Estimate	**Low**	**50% Low**	**Baseline**	**50% High**	**High**
Harvesting and Maintenance	na	$35/ton	$37.50/ton	$40/ton	$42.50/ton	$45/ton
Storage	$2 to $17/ton	$0/ton	$5/ton	$10/ton	$15/ton	$20/ton
Chipping	na	$8/ton	$9/ton	$10/ton	$11/ton	$12/ton
Transportation	$0.09 to $0.63 /ton/mile for 22 to 67 miles	$0.40/ton/ mile for 40 miles= $16/ton	$0.45/ton/mile for 45 miles=$20.25/ton	$0.50/ton/ mile for 50 miles= $25/ton	$0.55/ton/ mile for 60 miles= $33/ton	$0.60/ton/ mile for 70 miles= $42/ton
Stumpage	na	$0/ton	$0/ton	$0/ton	$2.50/ton	$5/ton
Total Cost ($/ton)		**$59**	**$72**	**$85**	**$104**	**$124**

Source: "Biomass Resources for Liquid Transportation Fuels," Chapter 2, NAS, *Liquid Transportation Fuels from Coal to Biomass* (2009). Notes: na = not available.

Supply Logistics

Low Energy Density Concerns

Some biomass feedstocks, such as agricultural residues and perennial grass feedstocks, have relatively low energy density, which can make transportation more costly, because a limited number of bales can be made to fit on a load. Because trucks become physically full when hauling biomass before reaching their weight limits, more truckloads would be required for cellulosic biofuel production relative to corn-based ethanol, thus, making the per-unit cost much higher for cellulosic feedstocks then for corn. In addition the greater number of trucks required to transport cellulosic feedstocks might also impose a heavy burden on rural transportation infrastructure.[54] Processing biomass before it goes to the plant will make it denser, and less costly to transport by allowing more biomass per load (especially in the case of round bales), and easier to handle once it arrives at the plant.[55] However, densification costs have been estimated at about $23 per dry ton, which may outweigh any savings related to transportation, especially if densification cannot take place immediately after harvest.[56] Densification at the storage facility would require the biomass to be transported as bales from the field and then stored until densification could take place.

Low energy density also means that a large volume of biomass would be needed by a plant to meet demand. For example, a 50 million gallon per year cellulosic ethanol plant would need a 90- feet-high pile of corn stover covering 100 acres of land in order to stay operational for a year.[57] Using a larger quantity of feedstock to make the same amount of fuel can make storage, transportation, and just in time delivery to plants more difficult and less efficient.

Harvest Timing Concerns

The timing of biomass harvest for particular feedstocks can have logistical advantages and disadvantages. For example, short-rotation woody crops can be harvested year around, which helps ease the burden of storage by being able to leave the biomass on the stump until it is needed. However, using the stump as storage will require year round harvest, which may be difficult should weather or equipment failures delay any regularly scheduled harvests. In contrast, corn stover should be harvested within three weeks after corn is harvested and before any snow might cover the field, which gives it a harvest window of about five weeks to three months depending on location. [58] This may require additional equipment and labor so as to not interfere with the corn harvest. Perennial grasses, on the other hand, can have harvest windows of up to 8 months depending on location.[59] *Miscanthus* must be harvested after the first frost in order to take advantage of the nitrogen retention through the root system.[60]

Because corn stover harvest is concentrated during one part of the year, long term storage will be needed to hold the corn stover until the plant is ready to use it (keeping in mind that biofuels plants are designed to run year round). For plants that can eventually use multiple feedstocks, they will be able to use corn stover around the time it is harvested, followed by feedstocks that become available at other times of the year.

Storage Concerns

Storage infrastructure for more dense crops including grains and oilseeds is available on many larger farms and at centralized elevator facilities. Large quantities of corn, for example, can be dried to a certain moisture content and stored for an extended period of time in a relatively small space. No such storage system or infrastructure currently exists for cellulosic biomass.

Forest biomass is attractive from a storage perspective, because it can be stored on the stump until needed, and depending on the climate and geography of its location, it can be harvested year round.[61] However, with the harvest of agricultural residues and perennial grasses being somewhat dependent on timing in the season, harvest cannot coincide with demand for biomass, which will require transportation and delivery to be tailored to the needs of the plant.

Between harvest and delivery, on-site storage of the biomass will be necessary. Storage conditions will vary from producer to producer and from region to region depending on weather conditions. While being stored, especially in outdoor condition, biomass can lose dry matter. Avoiding dry matter loss while storing biomass outside will depend on humidity and precipitation levels. Individual producers will likely be expected to store their own biomass until the plant needs it, which could be several months after it is harvested. With no central storage facility and producers not likely to have storage specifically to keep the feedstock until the plant needs it, biomass may be stored outside on the ground.

For biomass that is baled, the packaging method, whether it is with twine, net wrap, or plastic wrap can affect the amount of dry matter loss. Ideally, corn stover bales should at least be stored on a site with good drainage, a concrete or gravel surface, and under cover or inside if not wrapped with plastic.[62] Depending on the conditions, assumptions regarding dry matter loss, where biomass is stored, and the form in which it is stored, storage costs will vary. A study focused on corn stover found that storage adds between $7 and $13 per dry ton. [63]

For baled biomass, the choice between round bales and square bales depends on several conditions. One study found that round bales require less expensive equipment and a lower power tractor. They can be left in the field after harvest and brought to the storage facility later. Their round tops allow water to run off, which helps decrease dry matter loss. Round bales can be stored up to six months in satellite storage locations with less than 5% loss. Cost of storage is $2 per ton for round bales compared to $8 per ton for square bales[64] In contrast, another study found that using rectangular bales will reduce harvest and storage costs as the rectangular baler is able to bale more per hour that a round baler. However, rectangular bales have increased dry matter losses compared to round bales, so that the optimal solution may be a combination of different harvest and storage techniques.[65] **Table 14** shows estimated dry matter loss for switchgrass based on the type of bale, how it is covered, and the length of time it is stored. Each type of bale was stored on well-drained ground, gravel, and wooden pallets.[66] Overall, it is suggested that for biomass processed immediately after harvest, rectangular bales should be used, while for biomass that must be stored for more than three months, round bales with tarps should be used.

Research is being conducted to develop storage systems that keep the biomass wet while it is in storage. Biomass is mixed with additional water and made into large piles. Depending on the size of the pile, dry matter loss could be less than 5%. Advantages of a system like this include storing a given amount of biomass in a smaller area, softening biomass so as to weaken the lignin structure, and reduced risks of fire. It may also be more energy efficient to add water than to remove it. However, as biomass is stored in such wet conditions, the bacteria that forms can consume some of the sugar that would otherwise be fermented.[67]

Table 14. Switchgrass Dry Matter Loss by Bale Type and Cover System

Bale Type	Cover System	100 days	200 days	300 days	400 days
Round	None	6.0%	15.7%	14.0%	9.7%
Round	Tarp	0.0%	6.1%	4.6%	7.0%
Rectangular	None	27.2%	52.5%	52.1%	64.8%
Rectangular	Tarp	25.7%	20.8%	12.5%	13.7%

Source: English and Mooney (2009).

Moisture Content Concerns

Corn stover and switchgrass are assumed to have a moisture content of around 1 5%.[68] Short rotation woody crops, like poplar and willow, have about 50% moisture content at harvest.[69] In order for wood to be used for fuel, the moisture content cannot be above 65%. At this point, the energy required to dry the wood is greater than the energy content of the dry wood. The moisture content of forest residues tends to be between 40 and 60%, while the moisture content of primary mill residues, which include pulp, paper, and lumber, is around 20%, and the moisture content of secondary mill residues, which are kiln dried, is less than 10%.[70]

Quality Uniformity Concerns

Biomass can have highly inconsistent quality and characteristics, not only among feedstocks but within a feedstock type. With biomass feedstocks coming from numerous sources and producers over the course of an entire year for use at a single plant, it is initially processed so that the size will be as uniform as possible to optimize the conversion process. Plants receiving biomass that has been harvested differently and stored under a variety of conditions will incur additional costs to arrive at a uniform feedstock product.

Having a uniform feedstock delivered to the plant will decrease processing and pretreatment costs. With more experience and the development of the appropriate harvesting, collecting, and storage technologies, the pre-treatment processing step could take place before the biomass even arrives at the plant, which would eliminate the extra step and allow the plant to start with a uniform and consistent product, regardless of its source. In the case of corn stover, some processing may take place during harvest. As previously discussed, single-pass harvesting can allow stover to be chopped rather than baled. Stover in this form, however, will have high moisture content and may not be compacted.

Table 15. Estimated Farm Gate Cellulosic Feedstock Costs

Feedstock	Approximate Cost per Dry Ton
Switchgrass	$65 to $85
Miscanthus	$60 to $80
Corn Stover	$50 to $70
Short-Rotation Woody Crops	$50 to $60
Forest Residues	$45

Source: Estimates by Tyner, Brechbill, and Perkis, Purdue University, August 2010.

CONCLUSIONS

The production of cellulosic feedstocks is a challenge to the development of the cellulosic biofuels industry. The establishment, maintenance, harvest, storage, and transport of cellulosic feedstocks remain far from perfect and will need to be improved in order to reduce feedstock costs. Whether the feedstock is a residue or dedicated crop, there is not much experience among producers. **Table 15** includes best-guess estimates of farm-gate feedstock costs. Transportation would also need to be added to determine the delivered cost of feedstocks, but this will depend on how far feedstocks are from the plant location. Each feedstock option is faced with opportunities to reduce production costs, and the choice of which feedstock to use will depend upon the location of the plant and the local feedstock supply availability.

4. CELLULOSIC BIOFUEL CONVERSION TECHNOLOGIES

Biochemical and thermochemical conversion of lignocellulosic feedstocks are currently receiving the most attention as processing methods for cellulosic biofuels. Neither conversion process is ready for commercialization. Much of what is appearing is taking place in laboratories and demonstration and pilot plants that are small in scale. Plans for commercial plants have been announced by several companies. However, commercialization is likely to be slow to develop absent significant incentives, and cost reductions will be necessary for it to develop.

This chapter will focus on the biochemical and thermochemical conversion processes. Each process is described in some detail, and the likely areas for efficiency improvements and cost reductions are discussed. Energy yields and production costs for each process from recent studies is then presented. The current state of operational and proposed cellulosic biofuels plants is then outlined, along with a discussion of the funding thus far allocated by the federal government. Finally, there is an overview of some other developing technologies that are receiving attention as research continues to improve the efficiency of conversion processes.

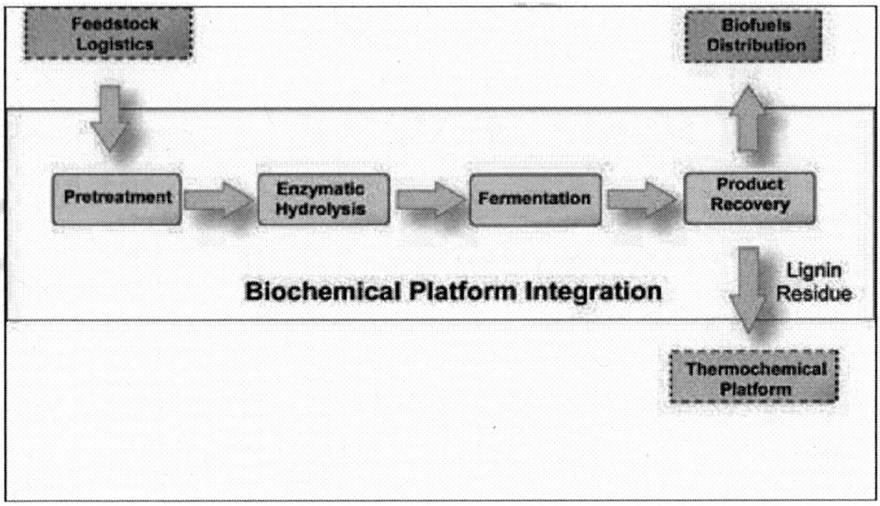

Source: U.S. Department of Energy, Energy Efficiency and Renewable Energy, *Biomass Program*, at http://www1.eere.energy

Figure 6. Biochemical Conversion Process

Biochemical Conversion

The biochemical conversion process, as summarized in **Figure 6**, has attracted the most attention so far because of its similarities to the process currently used to produce ethanol from corn grain.

Enzymes or acids are used to break down the plant into sugars that are then fermented into liquid fuel. Several processes are currently being researched and developed in laboratories, but it is difficult to know with any certainty whether those that appear successful in trials will also be successful on a commercial scale.

Pretreatment

The feedstock is pretreated by changing its chemical makeup in order to separate the cellulose and hemicellulose from the lignin (**Figure 7**). With the biochemical conversion process, the lignin cannot be fermented into liquid fuel, but it can be recovered for later use.

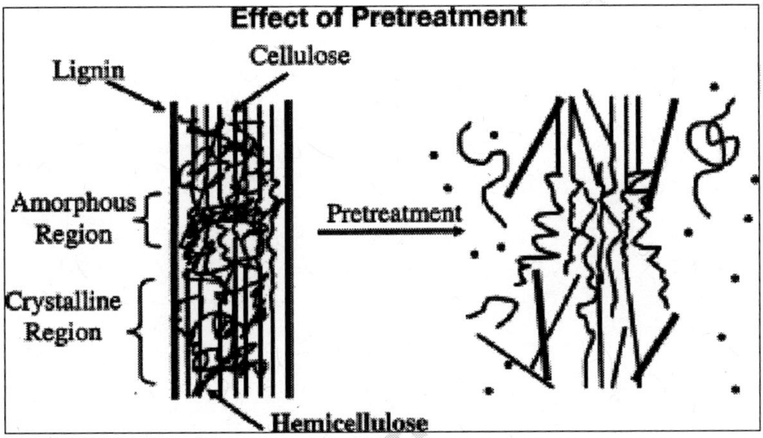

Source: Mosier, N., C. Wyman, B. Dale, R. Elander, Y.Y. Lee, M. Holtzapple, and M. Ladisch, "Features of promising technologies for pretreatment of lignocellulosic biomass," *Bioresource Technology* 96(6), 2005, pp. 673- 686.

Figure 7. Simplified Impact of Pretreatment on Biomass

Reducing the lignin content of biomass or modifying it in some way are alternatives that would reduce the inputs necessary for pretreatment and enhance the effectiveness of hydrolysis. A pretreatment system should maximize the remaining amount of sugar and keep the overall process as simple as possible in order to reduce costs relative to the amount of sugar that is recovered. It is ideal for a pretreatment process to be flexible enough to handle multiple types of biomass feedstock and to reduce the preparation and processing necessary. That is, the pretreatment processes should be able to utilize many different feedstocks in a variety of forms. Pretreatment should also minimize the amount of enzymes that must be used in hydrolysis. It is estimated that pretreatment of biomass accounts for 17% of capital costs.[71]

Biological pretreatment can use fungi to change the chemical composition of the feedstock by attacking the lignin. This process, however, can take 10 to 14 days, so it is not being focused on for commercialization. Physical pretreatment will perform a mechanical breakdown of the crystalline structure to increase the surface area and make it more susceptible to attacks by enzymes or acids in later steps. The lignin is not removed with this pretreatment, and the energy inputs are presently too high for commercialization. Chemical pretreatment is used in commercial pulping processes with the goal of maintaining the overall structure of the high value pulp. With biomass, the structure should be broken down, and the pretreatment process should be less expensive. Combination pretreatments that use both the physical and chemical pretreatment processes are being explored as an improved alternative for biomass.[72]

Hydrolysis

After sugars from the cellulose and hemicelluloses are separated and exposed during pretreatment, hydrolysis uses enzymes or acids to break down the complex chains of sugar molecules into simple sugars in preparation for fermentation. Hydrolysis will result in both glucose and xylose, which are six and five carbon sugars, respectively. Currently, enzymatic hydrolysis may be the better economic choice for commercialization relative to acid hydrolysis. Compared to acid hydrolysis, enzymatic hydrolysis is faster, more efficient, results in better yields, and uses less chemical input. Using enzymes in the hydrolysis of starch only requires one family of enzymes or cellulases. However, lignocellulosic biomass requires different cellulases to address the multiple parts of the plant.

The efficiency of hydrolysis is highly dependent on the effectiveness of pretreatment. If pretreatment leaves behind a large amount of lignin, the yield of simple sugars from hydrolysis may be reduced. The presence of lignin prevents enzymes from hydrolyzing. It is unlikely that the lignin can be completely removed from biomass, but there may be a possibility for changing some lignin characteristics in order to make it more compatible with the hydrolysis process. In the meantime, new types of catalysts that are able to deal with the lignin are being researched.

Hydrolysis can be an expensive step depending on the cost of the enzymes. These costs have definitely decreased over time but in many cases cost reduction is still a hurdle to commercialization. For a 50 million gallon per year plant, enzymes cost about 50 cents per gallon in 2009. By 2015, this cost is expected to be 44 cents per gallon.[73]

Fermentation

After pretreatment and hydrolysis have released simple sugars, fermentation is used to turn as much of that sugar as possible into liquid fuel. Hexoses, or six carbon sugars in the form of glucose, can be fermented using traditional yeast strains. Pentoses, or five carbon sugars in the form of xylose, are not fermented as easily, and research is being conducted to improve and develop microorganisms that can ferment pentose sugars. A yeast microorganism developed at Purdue University can ferment both pentoses and hexoses, which increases the ethanol yield by about 40%, and reduces costs by not requiring the sugars to be separated before fermentation.[74]

Distillation

Distillation occurs after the liquid fuel has been fermented in order to achieve a 95% pure form. Distillation is a well-established technology as it is used in corn based ethanol production.

Use of Lignin

Recovered lignin can be burned to generate electricity and steam to power the bio-refinery or for other outside uses. Research is being conducted to determine other ways in which the amount of lignin present in biomass can be reduced, which would help reduce the need for pretreatment.[75]

Improvements

The primary areas for potential improvement of the biochemical process are in pretreatment, hydrolysis, and fermentation. The effectiveness of pretreatment must be improved to better prepare the feedstock for hydrolysis. Enzyme costs must be reduced, and enzymes must be made more efficient. To do this, research is needed to identify and understand the inhibitors that block the breakdown of the biomass into simple sugars. Determining why lignin can resist enzymes and finding alternative uses for it will make the hydrolysis step more efficient. For fermentation, a single microorganism is needed that will ferment both five and six carbon sugars in one step. Currently, microorganisms are available to ferment each type of sugar in separate steps, but their effectiveness in an industrial capacity must still be proven. Ultimately, pretreatment, hydrolysis, and fermentation should be combined into as few steps as possible to reduce costs.

Thermochemical Conversion

Thermochemical conversion processes, which use heat to decompose the feedstock, are well established and developed. There are two main types of processes – gasification and pyrolysis. Gasification has been used primarily for converting coal into liquid fuels. The technology can also be used to convert biomass to liquid fuels. Using gasification, synthesis gas, or syngas, is produced and cleaned. The Fischer-Tropsch process is then used to convert the gas into liquid fuels of a variety of types. Using pyrolysis, a liquid bio-oil is produced and either used directly as fuel or converted into other types of fuel. Unlike biochemical conversion, thermochemical conversion uses the entire biomass including the lignin portion.

Gasification and Fischer-Tropsch Synthesis

Prior to gasification, the biomass must be 20% or less moisture. For practically any type of biomass, drying will be required. Established drying technologies usually take place at high temperatures, which creates an opportunity for improvement. Research is being done to dry biomass at lower temperatures and use excess heat from drying for other purposes.

Gasification, as summarized in **Figure 8**, is an anaerobic process where the partial combustion of biomass feedstock at over 700°C generates synthesis gas, a mixture of carbon monoxide and hydrogen.[76] Despite this being a technology already in use, producing synthesis gas from a biomass feedstock still requires improvement in cleaning up tar, ash, and other impurities in the synthesis gas that may disrupt the Fischer-Tropsch process of creating a liquid fuel by inactivating the catalyst. Fischer-Tropsch synthesis dates back to the 1920s and was used extensively by the Germans in World War II to create liquid fuel from coal.[77] Despite the longevity of the process, gasification to create a liquid fuel still requires further development to be commercially viable. The use of Fischer-Tropsch synthesis for liquid fuels remains on the back burner for most oil companies today and would only be used should the price of gasoline become extremely expensive.[78] Much of the current gasification technology available is used to generate power, which is similar to producing liquid fuels with respect to the catalyst used but is complicated by meeting purity standards associated with the use of biomass.

Pyrolysis

Pyrolysis, as summarized in **Figure 9**, is the partial combustion of biomass feedstock at 450°C to 600°C in the presence of no oxygen, which produces bio-oil.[79] Bio-oil, which is rich in carbon, is similar to crude oil and must be refined into biofuels. Because pyrolysis converts the biomass into a liquid form, it is easier to store and transport. Fast pyrolysis requires higher temperatures than slow pyrolysis but occurs in about two seconds. Currently, fast pyrolysis is receiving the most attention as a viable conversion process. Keeping the pyrolysis oil stable long enough to transform the bio-oil into hydrocarbons is one of the major barriers in the pyrolysis pathway.

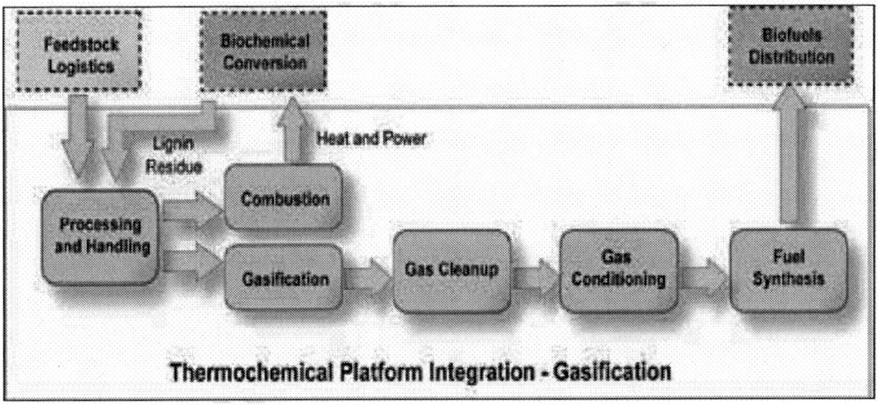

Source: U.S. Department of Energy, Energy Efficiency and Renewable Energy, Biomass Program, at http://www1.eere.energy

Figure 8. Thermochemical Conversion Process via Gasification

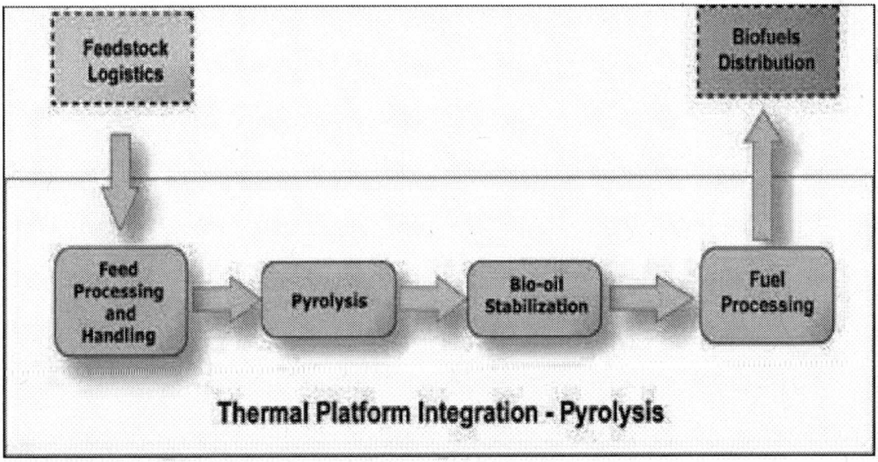

Source: U.S. Department of Energy, Energy Efficiency and Renewable Energy, Biomass Program, at http://www1.eere.energy

Figure 9. Thermochemical Conversion Process via Pyrolysis

Improvements

As mentioned above, thermochemical technology is already in use for the conversion of coal to gas for electricity generation. However, that technology is not currently being used to produce any liquid fuels. Gasification to produce synthesis gas also produces excess tar that is left in the gas. This excess tar requires that the synthesis gas be cleaned and conditioned. Stability of the bio-oil and cost are the major issues with pyrolysis.

Energy Yield

Another area for potential improvement relates to biofuel yield per ton of feedstock. **Table 16** outlines estimates for energy yields for both biochemical conversion and thermochemical conversion. The Department of Energy Biomass Program's Theoretical Ethanol Yield Calculator calculates the maximum theoretical yield with biochemical conversion of feedstocks based on their composition. Based on its results, the theoretical yields for corn stover, switchgrass, and forest thinning are 113, 97, and 82 gallons per dry ton, respectively (AFDC). As plants and technology approach commercialization, the rate of efficiency could approach 100% of the maximum theoretical energy yield.

Table 16. Energy Yields by Conversion Technology

Conversion Technology	Energy Yield	Expected Year
Biochemical (ethanol)	69.7 gal./ton 71.9 gal./ton 89.7 gal./ton 78.0 gal./ton 89.7 gal./ton	2007[a] 2007[b] 2007[a] 2008[c] 2015[c]
Thermochemical	80.1 gal./ton 47.0 gal./ton gasoline equivalent 61.4 gal./ton 94.1 gallons of mixed alcohols/ton	2007[e] 2008[g] 2009[h] 2015[g]

Source: Compiled by Tyner, Brechbill, and Perkis, Purdue University, August 2010, from sources listed.

Notes: The year associated with each study is the year that this energy yield is expected to be achieved.

a. Tiffany, D.G., *Economic Comparison of Ethanol Production from Corn Stover and Grain*, Agricultural Utilization Research Institute (AURI), Energy Users Conference, (Redwood Falls, MN), March 13, 2007.

b. Aden, A., *Biochemical Production of Ethanol from Corn Stover: 2007 State of Technology Model*, NREL/TP-5 10- 43205, National Renewable Energy Laboratory (Golden, CO), May 2008.

c. Bain, R.L., *World Biofuels Assessment, Worldwide Biomass Potential: Technology Characterizations*, NREL/MP-5 10- 42467, National Renewable Energy Laboratory (Golden, CO), December 2007, [hereafter referred to as Bain (2007)]; Foust, T.D., A. Aden, A. Dutta and S. Phillips, "An economic and environmental comparison of a biochemical and a thermochemical lignocellulosic ethanol conversion process," Cellulose, 1 6(4), 2009, pp. 547-565, [hereafter referred to as Foust et al (2009)]; and Tao, L. and A. Aden, "The economics of current and future biofuels," *Vitro Cellular and Developmental Biology*, 45(3), 2009, pp. 199-217, [hereafter referred to as Tao and Aden (2009)].
d. NAS, Liquid Transportation Fuels from Coal to Biomass (2009).
e. Rismiller and Tyner (2009).
f. Bain, R.L. (2007); Foust et al (2009); and Tao and Aden, (2009).
g. NAS, Liquid Transportation Fuels from Coal to Biomass (2009).
h. Wright, M.M. and R.C. Brown, "Comparative Economics of Biorefineries based on the Biochemical and Thermochemical Platforms," Biofuels, Bioproducts, and Biorefining 1 (2007), pp. 49-56, [hereafter referred to as Wright and Brown (2007)]; as taken from Rismiller and Tyner (2009).
i. Bain (2007) as taken from Rismiller and Tyner (2009).

Estimated Cost per Gallon

Considering that the production of cellulosic biofuels is in its infancy, predictions of the total cost are subject to a high degree of uncertainty. Most research in the area is focused on cost reduction, so up-to-date cost estimates can quickly become inaccurate with improvements to feedstocks and conversion technologies. **Table 17** outlines estimates from recent studies for total capital costs and operating costs per gallon for biochemical conversion.

Table 18 summarizes capital and operating costs for thermochemical conversion. The estimates vary based on the year they were conducted and the size of the plant. One study found that there is no distinct economic difference between biochemical and thermochemical conversion for cellulosic ethanol production.[80] The same study found that herbaceous feedstocks are better suited for biochemical conversion, while woody biomass is better suited for thermochemical conversion.[81]

Table 17. Biochemical Production Costs

Plant Size (MGPY)	Capital costs: $ Million ($/gal.)	Capital charge ($/gal.)	Operating costs ($/gal.) in gasoline equivalent: Feedstock, Energy, Enzyme, Other Variable Costs	Total Cost ($/gal.)	Data Year
25	$136 ($5.44)	$0.73	$1.50 x 1.5 = $2.25[a]	$2.98	1999[b]
45	$183 ($4.06)	$0.54	$1.48 x 1.5 = $2.22	$2.76	2007[c]
50	$338 ($6.76)	$0.91	$1.97	$2.88	2009[d]
69.3	$220 ($3.17)	$0.43	$1.33[e] x 1.5 = $2.00	$2.43	2007[f]
100	$349 ($3.49)	$0.79	$1.97	$2.76	2009[g]
150	$756 ($5.04)	$0.67	$1.76	$2.43	2005[h]

Source: Compiled by Tyner, Brechbill, and Perkis, Purdue University, August 2010, from listed sources.

Notes: MGPY = Million gallons per year. "Capital costs" per annual gallon are included in order to compare capital expenses for plants of various sizes. The "Capital charge" per gallon is estimated using an annual payment calculated at a 12% interest rate for a plant with a life of 20 years. The "Total Cost" per annual gallon equals the sum of the capital charge (per gallon) and operating costs (per gallon).

a. Prices published as $ per gallon of ethanol (volumetric basis) were multiplied by 1.5 in order to convert to $ per gallon of gasoline energy-equivalent basis.

b. McAloon, A., F. Taylor, W. Yee, K. Ibsen, and R. Wooley, *Determining the cost of producing ethanol from corn starch and lignocellulosic feedstocks*, NREL/TP-580-28893, National Renewable Energy Laboratory, (Golden, CO), October 2000.

c. Tao and Aden (2009)

d. Rismiller and Tyner (2009).

e. $1.33 per gallon is the minimum ethanol selling price.

f. Foust et al (2009).

g. NAS, *Liquid Transportation Fuels from Coal to Biomass* (2009).

h. Wright and Brown (2007).

Table 18. Thermochemical Production Costs

Plant size MGPY	Capital costs $ Million ($/gal.)		Capital charge ($/gal.)	Operating costs ($/gal.) in gasoline equivalent: Feedstock, Energy, Enzyme, Other Variable Costs	Total Cost ($/gal.)	Data Year
45	$241	($5.36)	$0.72	$1.32 x 1.5 = $1.98[a]	$2.70	2007[b]
45	$488	($10.84)	$1.45	$1.70	$3.15	2009[c]
61.8	$210	($3.17)	$0.42	$1.22d x 1.5 = $1.83	$2.28	2007[e]
67	$636	($9.49)	$1.27	$1.78	$3.05	2008[f]
150	$854	($5.69)	$0.76	$1.80	$2.56	2007[g]

Source: Compiled by Tyner, Brechbill, and Perkis, Purdue University, August 2010, from listed sources.

Notes: MGPY = Million gallons per year. "Capital costs" per annual gallon are included in order to compare capital expenses for plants of various sizes. The "Capital charge" per gallon is estimated using an annual payment calculated at a 12% interest rate for a plant with a life of 20 years. The "Total Cost" per annual gallon equals the sum of the capital charge (per gallon) and operating costs (per gallon).

a. Prices published as $ per gallon of ethanol (volumetric basis) were multiplied by 1.5 in order to convert to $ per gallon of gasoline energy-equivalent basis.

b. Tao and Aden (2009).

c. Rismiller and Tyner (2009).

d. $1.22 per gallon is the minimum ethanol selling price.

e. Foust et al, *Cellulose* (2009).

f. NAS, *Liquid Transportation Fuels from Coal to Biomass* (2009).

g. Wright and Brown (2007).

Current Plants

There are currently no commercial cellulosic biofuel plants in the United States, and plans for proposed plants are far from definite. However, as of April 2009, there were 25 operational pilot and demonstration cellulosic ethanol plants with approximately 3.5 million gallons of ethanol production capacity.[82] Of these plants, 17 are using biochemical conversion technology, seven are using thermochemical conversion technology, and one is using a combination of both technologies. These pilot and demonstration plants are using a wide range of feedstocks including corn stover, wheat straw, rice straw, sugarcane bagasse, switchgrass, wood residues, paper waste, and municipal solid waste. These plants are run by a combination of academic, government, and private organizations. Three plants are currently under construction and were expected to be operational by the end of 2009, but none achieved the objective.[83] These three plants, once operable, are expected to produce a combined total of over 10 million gallons of ethanol annually. Other planned or proposed plants are often quite small due to uncertainties regarding conversion technology and limits to funding.

According to *Ethanol Producer Magazine,* which maintains a list of proposed ethanol plants (include both corn and cellulosic plants in the United States and Canada), estimated that there were 118 proposed plants in 2008, but only 70 proposed plants in 2009.[84] Of those proposed in 2009, 30 planned to use cellulosic feedstocks and would produce just over one billion gallons per year if constructed according to proposed dimensions. However, the 40% decline in the number of proposed plants from 2008 to 2009, is indicative of how much uncertainty exists in predicting when future biofuels plants will begin construction, let alone when they will reach commercialization. It is likely that some of the decline is be attributed to difficulty securing financing in the fragile financial markets and to uncertainty in oil and corn prices.

Between 2002 and 2008, the Department of Energy's Energy Efficiency and Renewable Energy Biomass Program allocated more than $800 million in federal funding to both private companies and universities for advanced biofuels research and development. An additional $786.5 million from the American Recovery and Reinvestment Act (P.L. 111-5) has been slated to provide added funding for development of commercial sized bio-refineries.[85]

Notes: MGPY = million gallons per year. Mln = million.

Table 19 describes 10 plants that are receiving DOE or USDA support. In total, including these 10 plants, there are 16 planned cellulosic ethanol plants and 22 proposed cellulosic ethanol plants with a total annual production capacity of 639 million gallons.[86] Many of these plants are in the planning stages of conducting feasibility studies and securing funding and necessary permits. It is uncertain whether all of these plants will actually be constructed and become operational, but further research with DOE funded projects should increase the chance of success. The six plants that initially received $385 million in DOE funding in February 2007 were originally expected to be operational by 2011.[87] However, several of the plants have either postponed or cancelled their plans to proceed with construction (including Alico, Iogen, and Mascoma).[88]

Table 19. Cellulosic Ethanol Plants Receiving DOE or USDA Support Company

Company	Location	Feedstocks	Production Capacity: MGPY	Estimated Operating Date	Conversion Technology	DOE Funding ($ Mln)	Federal Loan Guarantee
Abengoa Bioenergy Corp.	KS	Corn stover, wheat straw, milo stubble, switchgrass	11.4	2012	Biochemical	$76	
BlueFire Mecca, LLC	CA	Woodchips, grass cuttings, other yard waste	17.0	TBD	Biochemical	$40	DOE
Ecofin/Alltech	KY	Corn cobs	1.3	2010	Biochemical	$30	
ICM Inc.	MO	Corn fiber/stover, sorghum, switchgrass	1.5	2010	Biochemical	$30	
Mascoma Corp.	MI	Wood fiber	40.0	2012	Biochemical	$26	
Pacific Ethanol	OR	Wheat straw, wood chips, corn stover	2.7	TBD	Biochemical	$24	
POET	IA		25.0	2011	Biochemical	$80	DOE
Project Liberty		Corn cobs/fiber					
Range Fuels	GA	Wood waste, switchgrass	40.0	2011	Thermo-chemical	$76	USDA
RSE Pulp and Chemical	ME	Woody biomass	2.2	2010	Biochemical	$30	
Verenium Corp.	LA	Sugarcane bagasse, wood, energycane	1.5	Online	Biochemical	$10	

Source: U.S. EPA (2009), Table 1.5-35, p. 170.

Other Technologies

BP and DuPont have entered into a partnership to research and develop biobutanol (which has an energy content just a bit less than gasoline), rather than ethanol.[89] Under current technology, biobutanol is more expensive to produce than ethanol; however, it has several inherent characteristics that make it a preferred transportation fuel over ethanol. Both first generation and second generation feedstocks can be used to produce biobutanol, which can easily be blended with gasoline. It can be blended at higher concentrations to avoid problems with the blending wall that currently face ethanol production. Biobutanol is easier to distribute than ethanol (given the current gasoline pipeline infrastructure). It can likely be shipped in existing pipelines and is less likely to separate in the presence of water than ethanol. After a few years of research and development, the BP and DuPont partnership is currently working on a demonstration plant and looks to have its first commercial plant operating by 2013.[90]

Researchers at Purdue University have developed new thermochemical processes that involve adding externally produced hydrogen to either a pyrolysis process or gasification process to enable capturing more of the carbon content in the cellulosic biomass.[91] This could increase the efficiency of the entire conversion process by possibly producing three times as much biofuels from the same quantity of biomass. While this process would reduce the amount of land needed to produce a given volume of biofuels, a cost-effective source of hydrogen must be secured. Researchers are focusing on a carbon-neutral hydrogen source like nuclear, wind, or solar for the long term and natural gas for the short term.

Conclusions

In order for cellulosic biofuels to be commercialized, the cost per gallon for conversion must be reduced, regardless of the conversion technology used, and the entire process must be made more efficient by increasing the biofuel yield per ton of feedstock. Biochemical and thermochemical conversion technologies are receiving the most attention for cellulosic biofuels production. Most of the development of cellulosic conversion technology is happening on a laboratory scale or in small demonstration or pilot plants. For the biochemical process, the sugars in cellulosic feedstocks must be broken down and separated before being converted into biofuels. This complex and rigid structure presents a challenge for cellulosic biofuels that is not faced by corn- based ethanol. Enzyme costs appear to be an important limiting factor for biochemical conversion. For thermochemical conversion, the higher capital cost is the major barrier to overcome. New cellulosic plants are receiving funding from DOE and USDA, but feedstock and conversion cost reduction will be the ultimate test for whether the cellulosic biofuels industry will achieve commercialization and be able to meet standards set by the Renewable Fuel Standard for advanced biofuels.

5. Economics and Policy of Cellulosic Biofuels

This final chapter of this chapter will look at some of the main government policies that are likely to impact the progress of cellulosic biofuels. After a brief review of policy and market developments, there is a discussion of the blend wall, its impact on the growth of biofuels in the United States, and some possible solutions to the challenges of the blend wall. Then, the Renewable Fuel Standard and the Biomass Crop Assistance Program (BCAP) are summarized. This is followed by a comparative discussion of fixed versus variable biofuels subsidies . Finally, the potential effects of fixed and variable subsides are evaluated using a plant-level cost model with both biochemical and thermochemical conversion processes and stochastic input prices.

Introduction

The United States has been the world's leading producer of ethanol since 2005 when it surpassed Brazil.[92] The biofuels industry began in the United States with corn-based ethanol in the late 1970s, and grew slowly until the oil price run-up of 2005 accelerated industry investment and growth (**Figure 10**). From the beginning, ethanol blending has been subsidized in the form of a tax credit at a rate of between 40 and 60 cents per gallon. Other supportive policies have been added in recent years, but the blending tax credit has been paramount from the beginning.[93]

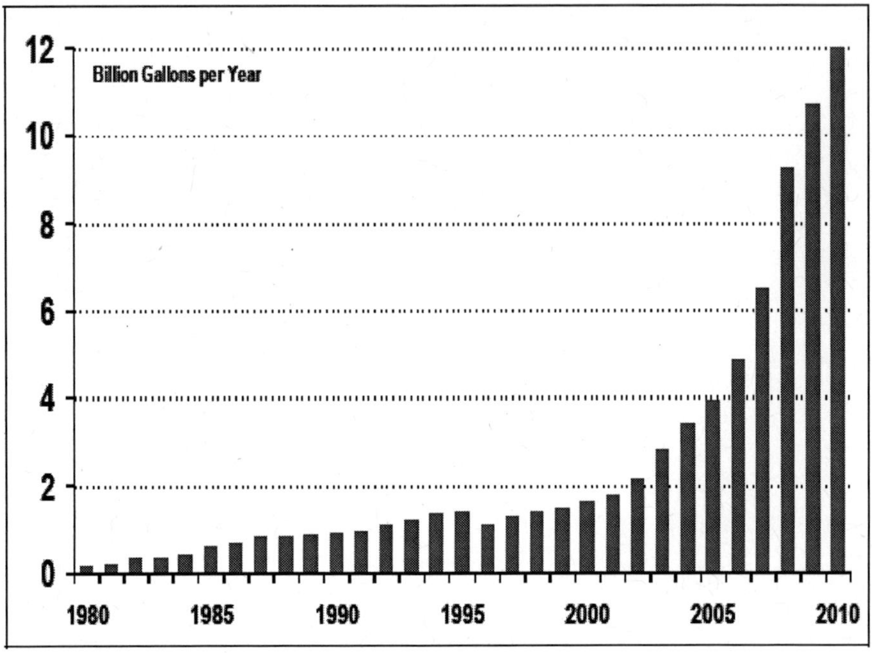

Source: 1980 to 2009, Renewable Fuel Association; 2010 is projected by the authors.

Figure 10. U.S. Ethanol Production from 1980 to 2010

Over the period 1983 to 2003, the price of oil averaged about $20 per barrel. When coupled with the blending tax credit and mandates for the use of oxygenates (including ethanol) in reformulated gasoline following the 1990 Clean Air Act Amendments, $20 oil was enough to promote slow, but steady growth of the U.S. corn ethanol industry.[94] Over this period, the price of ethanol varied but generally was equal to the gasoline price plus the government subsidy (**Figure 11**). That is, pricing was on a pure volumetric basis. Ethanol had value both as an oxygenate and for its higher octane level. Then in 2004 the price of oil began its move up, and the combination of a fixed subsidy keyed to $20 oil and much higher oil prices caused a boom in the corn ethanol industry. With high oil prices and substantial fixed subsidies, the ethanol industry enjoyed large profits and significant investment in new plants through 2007. The price relationship between ethanol and gasoline also became much more volatile as the quantity of ethanol on the market increased and other factors (such as the widespread phase out by May 2006 of MTBE, a major oxygenate competitor) came into play.

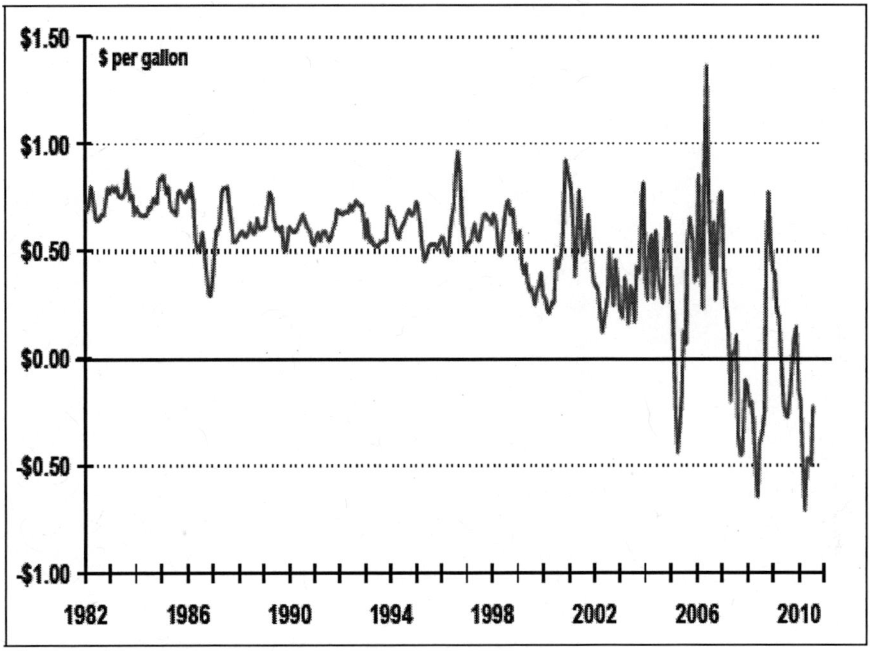

Source: Nebraska Ethanol Board, Lincoln, NE.

Notes: Prices are monthly average rack prices for ethanol and 87 octane, unleaded gasoline , Omaha, NE.

Figure 11. Historic Ethanol and Gasoline Price Differences

Beginning in 2008 the corn ethanol industry faced major economic difficulties as the price of corn surged. Corn is the single largest cost component in ethanol production accounting for nearly 66% of input costs.[95] Over two billion gallons of production capacity was shut down. Then even after corn prices fell, capacity remained down because the industry had reached the effective blend wall by mid-2009. The blend wall refers to the maximum quantity that can be blended given the 10% blending rate (more on this topic below). With excess supply on the market, the price of ethanol was ultimately forced down to the break-even price with corn. Conditions improved in late 2009 and early 2010 as more ethanol could be blended in winter months. However, in the spring of 2010, the industry returned to effectively operating at the blend wall. So long as the blending wall remains binding, this situation will persist.

The future of cellulosic biofuels and the potential to be profitable will depend a great deal on cost reductions in feedstock production and conversion technology, as discussed in the previous section. The economic situation and the policies implemented to support cellulosic biofuels will likely play a critical role in determining the path of commercialization just as they did for corn ethanol.

Blend Wall

The blend wall is a physical limit on the volume of ethanol that can be blended into the national transportation fuel supply due to the regulatory limit of 10% ethanol in fuel blends for standard vehicles.[96] As of August 2010, 10% was the national blending level. This blending limit was partially modified on October 13, 2010 (discussed below) to allow 15% blends for use in model 2007 and newer light-duty vehicles.

With annual U.S. gasoline consumption of about 140 billion gallons, the maximum amount of ethanol that can be blended to make E10 (90% gasoline and 10% ethanol) is 14 billion gallons.[97] To achieve 14 billion gallons of ethanol would require that every gallon of gasoline used across the country be blended at a 10% level. However, it is not possible to blend at that level everywhere. California has had a lower blending limit. In the South, blending has not been common in summer months because of higher evaporative emissions. There are problems with distribution to some regions. A more realistic estimate of the current blend wall is around 12 billion gallons or 9% of gasoline consumption.[98] **Figure 12** shows how U.S. ethanol consumption is converging with the amount of ethanol that would be needed if all gasoline were blended at a rate of 10%. This chart shows that gasoline use has only increased slightly over the past 15 years, but ethanol consumption has increased dramatically, especially over the past five years.

More ethanol use is possible with an increase in E85 (15% gasoline and 85% ethanol) consumption. However, E85 pumps are in short supply throughout the country and will likely not increase without strong government intervention.[99] The number of flexible fuel vehicles (FFVs) in the United States capable of running on E85 is estimated at over 8 million as of mid-2010,[100] but this accounts for only about 3% of the 240 million cars, vans, and light trucks that make up the nation's vehicle fleet.[101] Furthermore, very few of these are running on E85 even part of the time. If all 8 million FFVs ran exclusively on E85, they would consume an additional 4.7 billion gallons of ethanol (assuming 12,000 miles/vehicle-year and 20 mpg on ethanol).

Currently, less than 0.5% of total U.S. gasoline consumption is in the form of E85.[102]

An option for solving the blend wall is to increase the rate at which ethanol is blended with gasoline in conventional vehicles to 15% or 20% to increase the total amount of ethanol that can be used in the gasoline consumed by the conventional vehicle fleet. DOE and EPA are conducting tests to see whether higher blends will be compatible with current vehicles that run on conventional gasoline.[103] There is concern that newer vehicles with warranties may see those warranties voided if higher blends are used. In 2010, Underwriter's Laboratories (UL) certified that current gasoline dispensing equipment can handle blends containing up to 15% ethanol.[104] However, UL also said that E1 5 would not be suitable because, if one assumes that E1 0 may have some 15% ethanol, then some E1 5 will almost certainly have higher than 15% ethanol which may prove harmful to existing distribution infrastructure.

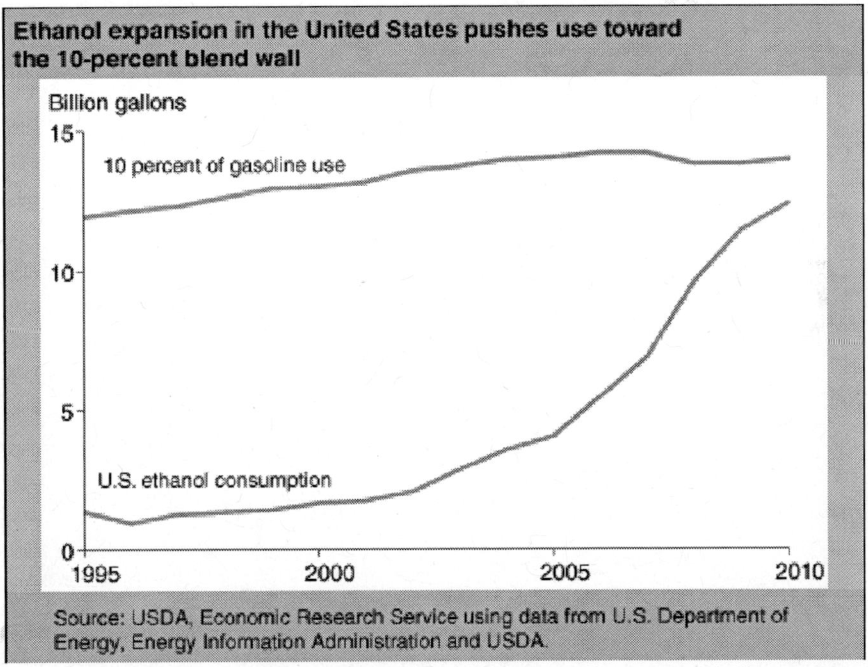

Source: Westcott, P.C., "Full Throttle U.S. Ethanol Expansion Faces Challenges Down the Road," *Amber Waves*, September 2009.

Figure 12. U.S. Ethanol Consumption and the 10% Blend Wall

In March 2009, Growth Energy, representing ethanol producers, applied for a waiver from the Environmental Protection Agency to increase the blending rate from 10% to 15%. On October 13, 2010, EPA issued a partial waiver for gasoline that contains up to a 15% ethanol blend (E1 5) for use in model year 2007 or newer light-duty motor vehicles (i.e., passenger cars, light-duty trucks, and sport utility vehicles). [105] A decision on the use of E1 5 in model year 2001 to 2006 vehicles will be made after EPA receives the results of additional DOE testing, possibly as early as November 2010. However, EPA also announced that no waiver would be granted for E15 use in model year 2001 and older light-duty motor vehicles, as well as in any motorcycles, heavy duty vehicles, or non-road engines. In addition to the EPA waiver announcement, numerous other changes have to occur before gas stations will begin selling E15 including many approvals by states and significant infrastructure changes (e.g., labeling of pumps, storage tanks, etc.). As a result, the vehicle limitation to newer models, coupled with infrastructure issues, are likely to limit rapid expansion of blending rates.

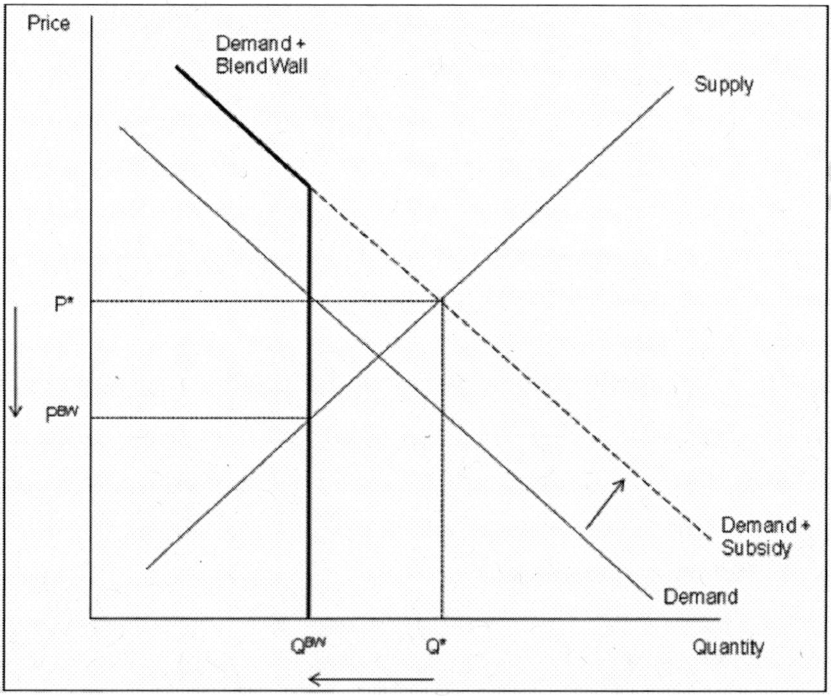

Source: Tyner, Brechbill, and Perkis, Purdue University, August 2010.

Figure 13. Ethanol Subsidies and the Blend Wall

Discussions of the blend wall have applied largely to the already established corn ethanol industry. However, with the eventual onset of commercial cellulosic biofuel production and a Renewable Fuel Standard that will require biofuels made from feedstocks other than corn, the issue of the blend wall and the current 10% blending rate may serve to hinder or even stop the development of the cellulosic biofuels industry before it gets started.[106] With mandated biofuels amounts higher than what can possibly be blended, the RFS may need to be capped and revised from its legislated levels.[107]

In effect, the blend wall is a constraint on the amount of ethanol that can be absorbed by the market. The presence of the blend wall serves to make the demand curve vertical, such that the same quantity of ethanol is demanded regardless of the price, once the blend wall quantity is achieved. The blend wall limits demand of ethanol to an amount that is less than the equilibrium amount (where ethanol supply equals unconstrained ethanol demand). This creates a market with reduced ethanol demand, excess ethanol supply, and an artificially low ethanol price and explains why some of the U.S. corn-based ethanol capacity was shut down in 2008 and 2009, and even through 20 10.[108]

Figure 13 shows the effect on the ethanol market of a blend wall. The fixed subsidy is also shown through the shift to the right of the demand curve. With a subsidy, more ethanol will be demanded at any given price. In **Figure 13**, the intersection between the supply curve and the demand curve with the subsidy indicates the quantity and price (P *) that would result without the blend wall. However, when the blend wall is in effect, the demand curve becomes vertical once that quantity (Q^{BW}) is reached. This demand curve with the blend wall is the bold line in **Figure 13**. The outcome will then be the intersection between the supply curve and the demand curve with the blend wall. Relative to the case where the blend wall is not in effect, the amount of ethanol consumed will decrease to the blend wall (Q^{BW}) and the ethanol price (P^{BW}) will also decrease.

If EPA determines to maintain the blend level at 10%, it will not be possible for cellulose ethanol to become established. Corn ethanol will be less expensive under most foreseeable conditions, so there would not be investment in cellulose ethanol even with the cellulose-based RFS. The perceived uncertainty would be too large to elicit industry investment. However, cellulosic-based biofuels produced via thermochemical conversion (e.g., bio-butanol) could be developed because they are not affected by the blend wall. These synthetic petroleum products can be used by existing petroleum-based distribution and storage infrastructure and the current fleet of U.S. vehicles.

However, even if the blending limit is increased to 15%, there still may not be room for much cellulose ethanol. From the description of the RFS below, one can see that in 2022 the corn (or conventional) component is 15 billion gallons, and the other "advanced biofuels" part is 4 billion. Sugarcane ethanol is included in the other advanced category. So corn plus sugarcane ethanol could sum to as much as 19 billion gallons. As noted above, 19 billion gallons is the effective blend wall at 15% blending. Since corn and sugarcane ethanol are likely to always be less expensive to produce than cellulosic ethanol, they could crowd out cellulosic ethanol even at the 15% blending level. Again, synthetic hydrocarbons produced via the thermochemical pathway would not be affected by the blend wall.

Renewable Fuel Standard

The Renewable Fuel Standard (RFS) was first introduced as a part of the Energy Policy Act of 2005 (P.L. 109-58). It required that 4 billion gallons of renewable fuels be used starting in 2006, increasing to a 7.5-billion-gallon requirement in 2012. The original RFS was never binding, meaning that the market always produced a quantity larger than the level of the mandate. The RFS was increased dramatically in the Energy Independence and Security Act of 2007 (EISA; P.L. 110-140). In 2008, 9 billion gallons of renewable fuels were required, and this requirement increases to 36 billion in 2022. There are also requirements as to what fraction of the total mandate must come from various types of fuels. **Figure 14** outlines the timing and break down of the requirements for the RFS.

The primary feature of the updated RFS is that it limits the amount of biofuels that can be produced from corn and still qualify under the RFS, and requires greenhouse gas (GHG) emission reductions relative to emissions from conventional gasoline for RFS-qualifying biofuels. From 2010 until 2015, the corn-based biofuels cap only increases slightly under the RFS, and eventually levels off in 2015 at 15 billion gallons. Any production of corn-based biofuels beyond this level will not count towards the mandate. For any corn-based ethanol from new refineries to count, it must reduce GHG emissions by 20% compared to conventional gasoline. However, all plants in production or under construction as of December 2007 are grandfathered—and that is most, if not all, of the eventual 15-billion-gallons corn-starch RFS.

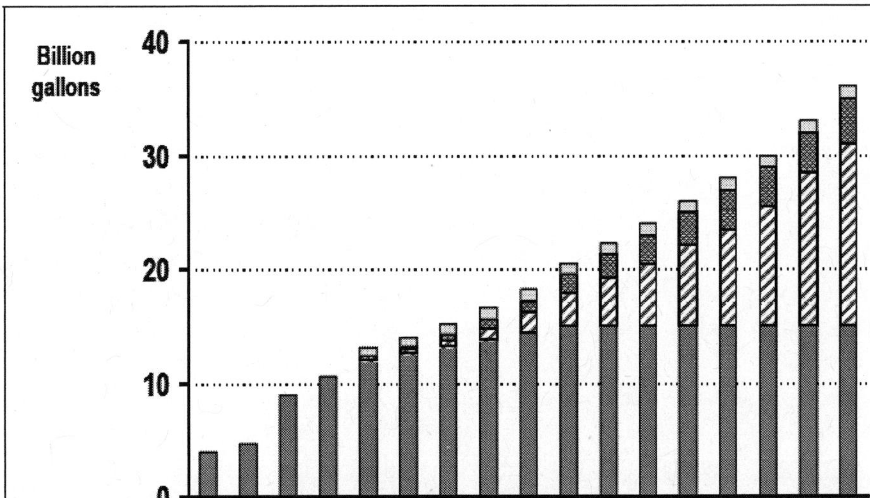

	2006	2007	2008	2009	2010	2011	2012	2013	2014	2015	2016	2017	2018	2019	2020	2021	2022
Biomass-based Diesel					0.7	0.8	1.0	1.0	1.0	1.0	1.0	1.0	1.0	1.0	1.0	1.0	1.0
Non-cellulosic Advanced					0.2	0.3	0.5	0.8	1.0	1.5	2.0	2.5	3.0	3.5	3.5	3.5	4.0
Cellulosic Advanced					0.1	0.3	0.5	1.0	1.8	3.0	4.3	5.5	7.0	8.5	10.5	13.5	16.0
Corn-starch Ethanol	4.0	4.7	9.0	10.5	12.0	12.6	13.2	13.8	14.4	15.0	15.0	15.0	15.0	15.0	15.0	15.0	15.0

Source: Energy Independence and Security Act of 2007, (EISA; P.L. 110-246)

Note: After 2012, the 1.0 billion gallons of biomass-based diesel is a lower limit. EPA can set the mandate higher, and reduce the "unspecified advanced biofuel" portion by an equal amount.

Figure 14. U.S. Renewable Fuel Standard (RFS) Mandates by Biofuel Type

In its final RFS ruling in 2010, EPA determined that corn-based ethanol from new plants meets the 20% GHG reduction criterion. The remaining fuels to meet the mandate must come from advanced biofuels that may be from either cellulosic or non-cellulosic sources. Advanced biofuels must be produced from a feedstock other than corn and achieve a 50% reduction in GHG emissions compared to conventional gasoline. Cellulosic advanced biofuels must be produced from any cellulose, hemicelluloses, or lignin from biomass and achieve a 60% reduction in GHG emissions compared to conventional gasoline. The emission reduction requirements may be reduced by 10% by the U.S. EPA should the initial reduction requirement not be feasible. [109] However, in its final 2010 ruling, EPA determined that all the biofuels it evaluated meet the GHG thresholds. Imported ethanol produced from sugarcane in Brazil, which is subject to approximately a $0.59 per gallon tariff (including a $0.54 most-favored nation duty and a 2.5% value-added tariff), may also be used for the other advanced biofuels category.

The RFS can be either non-binding or binding relative to the amount of biofuel that the market would otherwise demand. However, the two possible RFS scenarios generate very different market effects. The presence of the RFS will cause some portion of the demand curve to become vertical (as shown by the bold line in the following figures), and the effect of the RFS on the quantity and price will depend on how the RFS quantity compares to the quantity that would otherwise be demanded by the market. A non-binding RFS (**Figure 15**) will not affect the quantity or price, because the amount required by the RFS is below the market equilibrium. That is, the market is producing and consuming more ethanol than the RFS requires. However, a binding RFS (**Figure 16**) will have an effect on the quantity and price. When the RFS is binding, the market equilibrium quantity will be less than the RFS requires. Therefore, the RFS will cause the quantity of ethanol consumed and the ethanol price to increase.

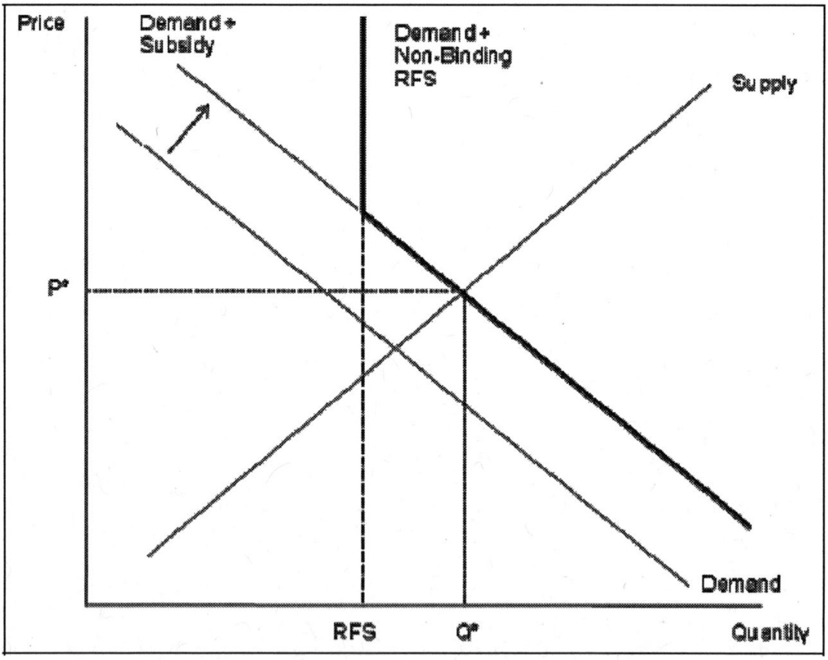

Source: Tyner, Brechbill, and Perkis, Purdue University, August 2010.

Figure 15. Ethanol Subsidies and Non-Binding RFS

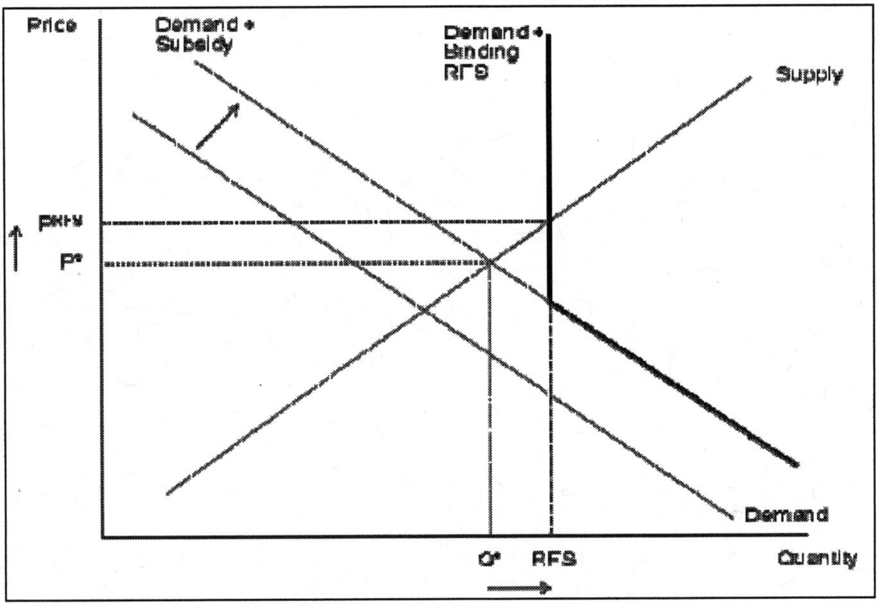

Source: Tyner, Brechbill, and Perkis, Purdue University, August 2010.

Figure 16. Ethanol Subsidies and Binding RFS

The government mandate on the use of biofuels creates demand and is intended to encourage investment in improved technology and new plants. Supporters of the RFS contend that the policy helps to reduce investment risk by guaranteeing demand over the years that the RFS is in place, increases energy security, reduces dependence on imported oil, and offers environmental benefits by requiring the use of renewable resources for energy production. Critics of the RFS view the policy as potentially hindering the development of other alternative energy technologies since it strictly promotes renewable fuels, while a technology neutral policy such as a cap-and-trade system or a carbon tax would promote the development of multiple renewable energy technologies.[110]

The RFS as it stands is based on renewable fuel use at specific levels rather than renewable fuel use as a share of overall fuel use. Therefore, improvements in fuel efficiency that have been mandated by the federal government may make meeting the RFS more of a challenge. With improved fuel efficiency, less fuel is necessary to travel a given distance. The Energy Information Administration forecasts U.S. motor gasoline consumption to fall from about 138 billion gallons in 2009 to 122 billion in 2022.[111] Therefore the

36 billion gallon RFS will rise from a 26% share to a 30% share of gasoline consumption on a volumetric basis.

In its final RFS ruling in 2010, the EPA effectively converted the RFS from a volumetric basis to an energy basis. EPA interpreted the RFS as 36 billion gallons of ethanol equivalent by 2022. A gallon of bio-gasoline, with 50% more energy than ethanol counts as 1.5 gallons of ethanol equivalent. The EU 2020 target is energy equivalent as well.[112] That is, 20% of the energy content of liquid fuels must be renewable. That approach is technology neutral with respect to renewable liquid transportation fuels in that it leaves it entirely up to the market place to determine the fuels that will be produced (subject to environmental constraints).

Biomass Crop Assistance Program

The Biomass Crop Assistance Program (BCAP) was established in the Food, Conservation, and Energy Act of 2008 (2008 farm bill; P.L. 110-246) to encourage biomass production by providing financial assistance with crop establishment, collection, harvest, storage, and transportation costs, as well as annual payments for biomass production, to producers within an economically practical distance from a biomass facility.[113] As of August 2010, an estimated total of $243 million has been paid out to BCAP participants.[114] Producers who sell biomass to qualified biomass conversion facilities can expect to receive up to 75% of the cost of establishing and planting eligible biomass crops.[115] Matching funds are expected to be available for collection, harvest, storage, and transportation to a biomass conversion facility. The maximum government payment is $45/ton.[116] Applications are being accepted for conversion facilities to qualify for the program. Plants that have qualified for the program are from a variety of geographic locations and are using a variety of biomass feedstocks. The Supplemental Appropriations Act of 2010 (P.L. 111-212) limits mandatory spending on BCAP by allowing no more than $552 million in FY2010 and $432 million in FY20 11; while no limit was placed on FY20 12 funding.[117]

Fixed Subsidies

Federal subsidies in the form of an ethanol tax credit were first introduced in the Energy Tax Act of 1978. The tax credit has remained between 40 and 60 cents per gallon since then.[118] Previously, this subsidy was a volumetric ethanol excise tax credit. In 2004, it was changed to a blender's tax credit, and is paid to the blender of the biofuel. The 2008 farm bill (P.L. 110-246) extended the blender's tax credit through 2010 but reduced it from $0.51 to $0.45 per gallon effective January 2009.

In addition to the $0.45 per gallon, there is a small ethanol producer credit of $0.10 per gallon for the first 15 million gallons produced each year by a small producer. Producers eligible for this credit must have an annual production capacity that is less than 60 million gallons.[119] About 1.6 billion gallons of ethanol applied for the small producer's tax credit in 2008. This credit is slated to last until the end of 2010.[120]

The 2008 farm bill also created a tax credit of $1.01 per gallon for producers of cellulosic biofuels.[121] Cellulosic ethanol that is also eligible for the volumetric ethanol excise tax credit, the small ethanol producer credit, or any other credit will only receive a total credit of $1.01 per gallon. This credit is set to expire at the end of 2012.

Variable Subsidies

The rapid increase in U.S. ethanol production (**Figure 10**) since 2005 has led to a concomitant increase in federal costs in the form of foregone revenue via the various tax credits (i.e., the fixed subsidies mentioned above). In 2009 the various fixed subsidies for biofuels in the United States cost nearly $5.9 billion, and is projected to reach at least $27 billion by 2022 if all tax credits are extended alongside of the RFS.[122] The biofuels tax credits have been available even during periods of extreme profitability such as occurred during 2006 and 2007. The high federal costs associated with the tax credits, coupled with the outlook for rapid growth in the cost of the fixed subsidy, have generated some interest among economists and federal budget watchdogs in favor of substituting a variable subsidy in place of the fixed subsidy as a means of lowering government costs while maintaining support for the biofuels industry.[123]

Unlike a fixed subsidy which remains constant in spite of market conditions, a variable subsidy adjusts its value depending on the current price of a barrel of oil. At high oil prices, above a certain threshold level, no subsidy is available, but as oil prices fall to the threshold level and then below, the variable subsidy kicks in and increases in value. Therefore, the subsidy amount paid is not constant as it is with a fixed subsidy.

In other words, a variable subsidy is designed to take into consideration the linkage between the ethanol industry's profitability and petroleum prices. In the absence of a blend wall, profitability of biofuel production will depend largely on the price of gasoline (which is highly correlated with oil prices), the primary commodity with which biofuels will compete. When oil and gasoline prices are high, biofuels are able to compete and remain profitable, while low oil prices make it difficult for biofuels to compete. Fixed subsidies ignore these differences. Fixed subsidies simply add to a plant's profitability, regardless of whether or not the plant would be profitable without the subsidy. In contrast, variable subsidies change as oil prices change. A variable subsidy can reduce overall government costs since large subsidy amounts will not be paid all the time, but only when the oil price warrants in order for biofuels to be competitive.

Variable subsidies could be linked to either oil or gasoline prices, but it is expected that the difference between the two would be minimal considering the strong statistical relationship between oil and gasoline prices.[124] In designing a variable subsidy, an oil price where subsidies start must be chosen. This is often referred to as a trigger price or a cut off price. The rate of change must also be determined, such that for every dollar that the oil price falls below the cut off oil price, the subsidy will increase by the rate of change. Variable subsidies can be calculated given the average oil price on a monthly or quarterly basis. Generally, the variable subsidy would provide a safety net for biofuels producers when oil prices are low and allows the market to drive biofuels production without government intervention when oil prices are high. A recent analysis of the variable subsidy with respect to corn-based ethanol production found that the variable subsidy provides a net present value (NPV) for the producer that is similar to that with the fixed subsidy, but that the variable subsidy could decrease risk and the probability of a loss on investment.[125]

Analytical Comparison of Fixed and Variable Tax Credits

The remainder of this chapter will evaluate the impacts on ethanol producing firms of a variable versus a fixed tax-credit subsidy. This comparison is made, not as a policy recommendation, but (in response to widespread news and media coverage) strictly as a comparative analysis to aid Congress' understanding of the differences between the two policy options.

Note on Volumetric Pricing (VP) versus Energy-Equivalent Pricing (EEP)

The price of ethanol can be compared to gasoline on either an energy-equivalent basis or a volumetric basis. The choice is crucial because the two methods lead to different results due to the lower energy content of ethanol—the energy from a gallon of ethanol (measured in Btu's) is only about 67% of the energy content of a gallon of gasoline. Thus, on a volumetric basis the price of a gallon of ethanol is deemed equivalent to the price of a gallon of gasoline, but on an energy-equivalent basis the ethanol price equals only about 67% of the gasoline price.

Historically, since ethanol has been blended at very low rates (2% to 3%) with gasoline their prices have been linked primarily on a volumetric basis (**Figure 11**). In the near term volumetric pricing might work for the low level blends such as E10 or E1 5. However, as the ethanol blending share increases in the future it will likely be linked more on an energy basis. For example, consider the 85% blending rate implied by E85. Since E85 is expected to get 22% lower mileage than E10 (based on energy content), presumably consumers would only be willing to pay for E85 about 78% of the price of E10.[126]

Thermochemically-produced biofuels are nearly energy-equivalent to petroleum products (unlike biochemically-produced biofuels with a 67%-equivalency rate). As a result, the following analysis will consider cases in which biofuels produced by thermochemical methods receive a subsidy 49% greater than ethanol when calculated on an energy-equivalent basis.

PROFITABILITY MODEL WITH UNCERTAINTY

A spreadsheet model is used to evaluate the impacts of a fixed versus a variable (tax credit) subsidy on grain and cellulosic biofuels under three different processing technologies: the traditional corn-starch-to-ethanol process, as well as both biochemical and thermochemical processing of cellulosic biomass. In addition to the different subsidy schemes and processing technologies, the spreadsheet model also incorporates risk and uncertainty with respect to oil prices, natural gas prices, corn prices, and biomass prices.[127]

The model assumes that the life of the biofuel plant will be 22 years total with 2 years for construction and 20 years of operation. The profitability of biochemical and thermochemical cellulosic biofuel production processes are compared to corn-based ethanol production by calculating the net present values (NPVs) per gallon.

For this analysis, some parameters from the original model (Rismiller and Tyner (2009)) have been made stochastic in order to incorporate risk and uncertainty. Stochasticity is incorporated using @Risk, an add-in to Excel that runs Monte Carlo simulations to choose input parameters within a specified distribution, and then uses the results of the calculations from each draw from input distributions to produce output distributions, rather than single values, for the NPV of plants using different biofuel production processes. In this case, a simulation with 10,000 iterations is run and a random draw from each input distribution is used to calculate the output value. Once the simulation is complete, the mean and standard deviation of the outputs are calculated based on the results from all iterations.

Below is a brief description of how each parameter was altered. Unless otherwise stated, all amounts are adjusted to 2010 real dollars.

Oil Prices: A simple mean-reverting forecast[128] is used to approximate the low, middle, and high forecasts of oil prices from the 2010 DOE *Annual Energy Outlook*.[129] Each set of forecasts has a deterministic mean-reverting component as well as a random component for each year. The price volatility introduced by the random component provides a testing ground for the stated benefits of the variable subsidy. The oil price forecasts are illustrated in **Figure 17**.

Natural Gas Prices: Moving into 2010, the link between natural gas prices and oil prices appeared to have become weaker as new sources of natural gas were discovered in the United States.[130] The natural gas price is a random draw each year and is not linked to the previous year's price. The mean of $5.34 and the standard deviation of $0.92 was based on the monthly industrial price for natural gas from June 2009 to May 2010 and was used to create the distribution from which the natural gas price was drawn. The minimum natural gas price is $3 per thousand cubic feet, and the maximum natural gas price is $15 per thousand cubic feet, which are close to the recent historical minimum and maximum.

Corn Prices: The corn price is specified as a function of the oil price given the link between energy and agricultural prices that has arisen in recent years.[131] When done this way, a different corn price will be used each year as a function of the stochastic oil price described above and a random component. The stochastic oil price for each year is plugged into the following regression of corn prices on oil prices:[132]

$$\textit{Corn price} = 1.78 + 0.029 * (\textit{oil price}) + eo$$

(where *eo* is a random normal error)

Regardless of how the corn price is calculated, the minimum corn price is $1.50 per bushel and the maximum corn price is $7.00 per bushel, which are close to the recent historical minimum and maximum.

DDGS Prices: The price of Distiller's Dried Grains and Solubles (DDGS) determines the revenue received by the corn-based ethanol plant for selling DDGS as a by-product. This price is calculated as a function of the corn price with the following equation:[133]

$$\textit{DDGS price} = 38.27 + 22.77 * (\textit{corn price}) + ec$$

(where *ec* is a random normal error)

Gasoline Price: The gasoline price is calculated as a function of the stochastic oil price:

$$\textit{Gasoline price} = 0.35 + 0.023 * (\text{oil price}) + eg$$

(where *eg* is a random normal error)

Using the stochastic oil price results in a different gasoline price each year.

Ethanol Price: Subsidies for ethanol blending and cellulosic biofuel production could be either fixed or variable. A fixed subsidy will be a per-gallon amount, while a variable subsidy will depend on a threshold oil price and a rate of change in the subsidy for each dollar below the threshold oil price that the market oil price falls. Current policy has cellulosic biofuel subsidies fixed on a volumetric basis ($1.01/gallon regardless of what biofuel is produced); for comparative purposes they will also be tested using an energy-equivalent basis.

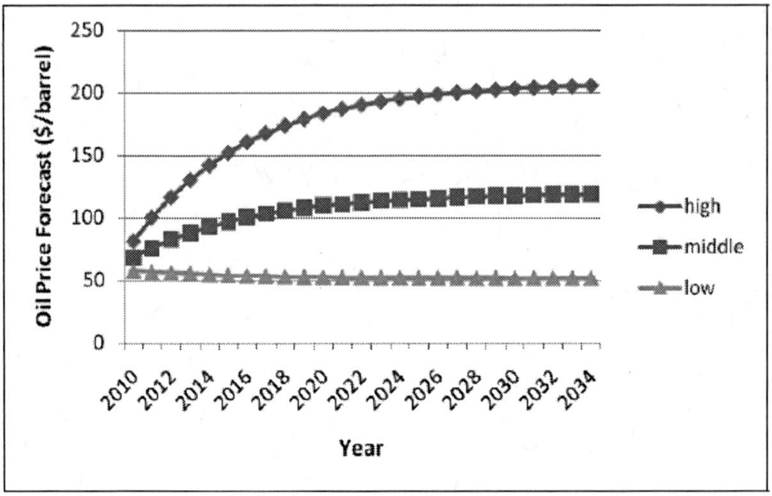

Source: Tyner, Brechbill, and Perkis, Purdue University, August 2010.

Figure 17. Mean Oil Price Forecasts for Stochastic Simulations (2008 real dollars)

Breakeven Oil Prices

The breakeven oil price represents the price for oil at which the plant's net present value (NPV) is zero. At higher oil prices the plant operates with a positive NPV (i.e., the plant operates at a profit); at lower oil prices the plant operates with a negative NPV (i.e., at a loss). Breakeven oil prices are calculated for both volumetric and energy-equivalent biofuels pricing methods (and with and without fixed subsidies) for all three processing technologies (**Table 20**).

When ethanol is priced on a volumetric basis relative to gasoline, grain-based ethanol and biochemical conversion have lower breakeven oil prices than thermochemical conversion, both with and without the fixed subsidy. When biofuels are priced on an energy-equivalent basis, the price of a gallon of ethanol will be 67% of the price of a gallon of gasoline, rather than equal to the price of gasoline as it is under volumetric pricing. As a result, on an energy-equivalent pricing basis the breakeven oil prices for grain-based ethanol and biochemical conversion increase, while the breakeven oil price for thermochemical remains unchanged. Thus, thermochemical conversion has a better chance of competing with grain-based and biochemical ethanol under energy-equivalent pricing.

Table 20. Breakeven Oil Prices ($/barrel)

	Grain	Biochemical	Thermochemical	
			VP	EEP
Volumetric Pricing (VP)				
Without Fixed Subsidies	$71.45	$126.68	$143.92	$143.92
With Fixed Subsidies[a]	$56.33	$92.74	$113.77	$98.92
Energy-Equivalent Pricing (EEP)				
Without Fixed Subsidies	$114.19	$196.64	$143.92	$143.92
With Fixed Subsidies	$91.62	$145.98	$113.77	$98.92

Source: Tyner, Brechbill, and Perkis, Purdue University, August 2010.

Note: VP = volumetric biofuels pricing; EEP = energy-equivalent biofuels pricing. The breakeven oil price is indicated for the deterministic case (i.e., without incorporating any of the stochastics described in the text). All commodities are assumed to be at current market prices except for gasoline and ethanol. Gasoline is priced based on the relationship provided in the text (without the random error) and ethanol prices are linked to gasoline by volume or energy equivalency.

a. The fixed subsidies are $0.45 per gallon and $1.01 per gallon for ethanol and cellulosic, respectively, based on current legislation.

With a fixed subsidy included when calculating the plant NPV, the breakeven oil price declines in all cases (assuming that the subsidy is fully passed on to the biofuel producer, which will not always be the case). Ethanol pricing decisions do not affect thermochemical conversion. However, an energy-equivalent subsidy which takes into account the thermochemical product's higher energy rating in comparison to ethanol would make the thermochemical product competitive with both ethanol options.

Profitability Results: Deterministic Case

The deterministic model is used to evaluate the profitability of biofuels production under both volumetric and energy-equivalent pricing methods but using current market prices for oil rather than the range of high, middle, and low from the previous section (**Table 21**). Grain and biochemical conversion do best when ethanol is priced on a volumetric basis relative to gasoline (primarily because they avoid the discount penalty related to lower energy content). However, only grain ethanol shows profitability even under volumetric pricing conditions.

Table 21. Profitability (NPV) with Fixed Subsidies, Deterministic Case

	Grain	Biochemical	Thermochemical	
			VP	EEP
$ per gallon of annual capacity				
Volumetric Pricing (VP)	$2.32	-$2.51	-$5.42	-$3.41
Energy-Equivalent Pricing (EEP)	-$1.58	-$6.42	-$5.42	-$3.41

Source: Tyner, Brechbill, and Perkis, Purdue University, August 2010.

Note: Negative values are in bold. VP = volumetric biofuels pricing; EEP = energy-equivalent biofuels pricing.

Table 22. Hypothetical Variable Biofuel Subsidy Under Various Oil Price Scenarios.

Market price of oil ($/barrel)	Variable Subsidy ($ per gallon of biofuel)
$110	$0.00
$100	$0.00
$90 (threshold)	$0.00
$85	$0.10
$80	$0.20
$75	$0.30
$70	$0.40
$60	$0.60

Source: Tyner, Brechbill, and Perkis, Purdue University, August 2010.

Subsidies given for biofuels made from cellulosic biomass are currently done on a volumetric basis. Whether cellulosic biomass is used to produce a gallon of ethanol through biochemical conversion or a gallon of gasoline through thermochemical conversion, both receive a fixed subsidy of $1.01 per gallon. It is uncertain whether pricing and subsidies will change in the future to an energy-equivalent basis.

Profitability Results: Stochastic Case

Variable Subsidy Defined

Stochastic model simulations are adopted to evaluate the variable subsidy whose per-unit value changes as market conditions change. A variable subsidy functions by providing firms with more support at low oil prices and withdrawing the subsidy at higher oil prices. The hypothetical variable subsidy evaluated in this analysis is based on a $90 per barrel oil price threshold and a $0.02 per gallon increment in additional subsidy for each dollar that the oil price is below the cut off. The following table (**Table 22**) shows variable subsidy rates at different oil market prices.

The net effect of replacing a fixed subsidy with a variable subsidy is that, while firms decrease their chances of operating at a loss, and government subsidies are reduced, there can be a slight drop in NPV compared to the fixed subsidy because no subsidy is paid out when the price of oil is $90 or higher. Thus, if profits for the various processing technologies are negative under a fixed subsidy across several potential oil price forecasts, then it is likely that the technology will not be profitable with a variable subsidy either. More research is needed to define a variable subsidy structure that works well for cellulosic biofuels.

Stochastic Baseline

To provide a baseline for profitability under stochastic conditions, **Table 23**, **Table 24**, and **Table 25** provide the stochastic means and standard deviations for the net present value (NPV), the probability of a loss, and the coefficient of variation with a fixed subsidy for each of the three oil price forecast scenarios: low, medium, and high oil price forecasts, respectively. Because different annual production capacities area assumed for each of the technologies (100 MGPY for corn ethanol and 50 MGPY otherwise), all NPV are expressed in per-gallons-of-annual-capacity to facilitate comparison.

In addition to NPV comparisons demonstrating profitability across different scenarios, these three tables include data to facilitate a comparison of the per-gallon-of-capacity **Subsidy Cost** (i.e., the tax revenue foregone by the tax credit), the new **Tax Revenue** (equal to the fuel tax revenue obtained from producing new biofuels), and the **Net Government Cost** (which equals the **Subsidy Cost** minus the **Tax Revenue**). Each government measure is also expressed as a NPV per gallon of annual capacity.

In general, all three of the technologies show positive profits with a high degree of probability when high oil prices are forecast, due primarily to higher oil-induced revenues. But profitability drops off at lower oil price forecasts, and as energy-equivalent pricing is adopted for biochemical processing and volumetric pricing for thermochemical processing. Note that the fixed subsidy cost shows no variability across the different oil price forecasts since payments are made for each gallon without respect to market conditions.

Table 23. Profitability with Fixed Subsidies at Low Oil Price Forecasts

	Grain		Biochemical		Thermochemical	
	VP	**EEP**	**VP**	**EEP**	**VP**	**EEP**
NPV ($/gallon of capacity)	$1.43	**-$1. 63**	**-$5.08**	**-$8.14**	**-$7.99**	**-$5.98**
(NPV standard deviation)	($1.16)	($0.58)	($2.45)	($2.03)	($2.60)	($2.61)
Probability of a Loss (%)	11.1%	99.7%-	98.4%	100.0%	100.0%-	98.9%
Coefficient of Variation	0.81	**0.36**	**-0.48**	**-0.25**	**0.33**	**-0.44**
Subsidy Cost	$2.68	$2.68	$6.01	$6.01	$6.01	$8.98
Tax Revenue (TR)	$0.59	**-$0.43**	**-$1.28**	**-$2.30**	**-$1.99**	**-$1.32**
(TR standard deviation)	($0.39)	($0.19)	($0.82)	($0.60)	($0.87)	($0.87)
Net Govt. Cost (NGC)	$2.08	$3.11	$7.29	$8.31	$8.01	$10.30
(NGC standard deviation)	($0.39)	($0.19)	($0.82)	($0.68)	($0.87)	($0.87)

Source: Tyner, Brechbill, and Perkis, Purdue University, August 2010.

Notes: All prices are expressed in per-gallon-of-annual-capacity to facilitate comparison across different plant sizes. NPV = Net Present Value. Negative values are in bold. VP = volumetric biofuels pricing; EEP = energy- equivalent biofuels pricing. The standard deviation is a statistical measure of deviation around the mean value. The Coefficient of Variation (CV) is the ratio of the standard deviation over the mean value. The CV is a useful statistic for comparing the degree of variation from one data series to another, even if the means are drastically different from each other.

Table 24. Profitability with Fixed Subsidies at Middle Oil Price Forecasts

	Grain		Biochemical		Thermochemical	
	VP	EEP	VP	EEP	VP	EEP
NPV ($/gallon of capacity)	$5.76	$0.42	$1.81	-$3.53	-$1.43	$0.58
(NPV standard deviation)	($2.78)	($1.39)	($4.56)	($3.26)	($4.58)	($4.55)
Probability of a Loss (%)	0.6%	40.9%	36.2%	85.9%	62.9%	46.4%
Coefficient of Variation	0.48	3.31	2.52	-0.92	-3.20	7.84
Subsidy Cost	$2.68	$2.68	$6.01	$6.01	$6.01	$8.98
Tax Revenue (TR)	$2.04	$0.26	$1.02	-$0.76	$0.20	$0.86
(TR standard deviation)	($0.93)	($0.46)	($1.52)	($1.09)	($1.53)	($1.52)
Net Govt. Cost (NGC)	$0.64	$2.42	$5.00	$6.78	$5.82	$8.11
(NGC standard deviation)	($0.93)	($0.46)	($1.52)	($1.09)	($1.53)	($1.52)

Source: Tyner, Brechbill, and Perkis, Purdue University, August 2010.
Notes: See notes for Table 23.

Under a scenario of low forecast oil prices and a fixed subsidy, only grain ethanol with volumetric pricing is profitable (**Table 23**). This is perhaps most representative of the biofuels industry's current situation since cash oil prices in the mid-$70s are presently closest to the levels of the low oil price forecast case.

As oil price forecasts increase, ethanol and biofuels priced to compete with oil and gasoline should command more revenue and become more profitable. The middle oil price forecast scenario confirms this (**Table 24**). Note however that the standard deviation in NPV does not allow for definitively ruling out negative profits in most cases (i.e., NPV minus its standard deviation < 0), even when NPV's are positive.

Only grain ethanol priced volumetrically has an NPV large enough to ensure profits in most cases (99.4% of the time). If oil prices follow the high-level forecast, most alternative biofuel technologies will be profitable with a high probability due to higher oil-induced revenues (**Table 25**).

Table 25. Profitability with Fixed Subsidies at High Oil Price Forecasts

	Grain		Biochemical		Thermochemical	
	VP	**EEP**	**VP**	**EEP**	**VP**	**EEP**
NPV ($/gallon of capacity)	$12.73	$4.48	$10.63	$2.40	$7.01	$9.04
(NPV standard deviation)	($5.61)	($3.29)	($7.15)	($4.89)	($7.09)	($7.02)
Probability of a Loss (%)	0.02%	6.3%	6.8%	32.2%	17.2%	10.0%
Coefficient of Variation	0.44	0.73	0.67	2.04	1.01	0.78
Subsidy Cost	$2.68	$2.68	$6.01	$6.01	$6.01	$8.98
Tax Revenue (TR)	$4.36	$1.61	$3.96	$1.21	$3.01	$3.69
(TR standard deviation)	($1.87)	($1.10)	($2.38)	($1.63)	($2.36)	($2.34)
Net Govt. Cost (NGC)	**-$1.68**	$1.07	$2.06	$4.80	$3.01	$5.29
(NGC standard deviation)	($1.87)	($1.10)	($2.38)	($1.63)	($2.36)	($2.34)

Source: Tyner, Brechbill, and Perkis, Purdue University, August 2010.
Notes: See notes for Table 23.

Table 26. Profitability with Volumetric Pricing for Grain Ethanol, Stochastic Case

	Low Oil	Middle Oil	High Oil
Fixed Subsidy			
NPV ($/gallon of capacity)	$1.43	$5.76	$12.73
(NPV standard deviation)	($1.16)	($2.78)	($5.61)
Probability of a Loss	11.1%	0.6%	0.02%
Coefficient of Variation	0.81	0.48	0.44
Variable Subsidy			
NPV ($/gallon of capacity)	$1.09	$4.44	$11.09
(NPV standard deviation)	($0.50)	($2.24)	($5.26)
Probability of a Loss	0.5%	0.0%	0.0%
Coefficient of Variation	0.46	0.50	0.47

Source: Tyner, Brechbill, and Perkis, Purdue University, August 2010.
Notes: See notes for Table 23.

Table 27. Subsidy Costs, Tax Revenues, and Net Government Costs with Volumetric Pricing for Grain Ethanol, Stochastic Case

	Low Oil	Middle Oil	High Oil
Fixed Subsidy		**NPV per gallon of capacity**	
Subsidy Cost	$2.68	$2.68	$2.68
Tax Revenue	$0.59 ($0.39)	$2.04 ($0.93)	$4.36 ($1.87)
Net Government Cost	$2.08 ($0.39)	$0.64 ($0.93)	**-$1.68** ($1.87)
Variable Subsidy		**NPV per gallon of capacity**	
Subsidy Cost	$2.23 ($1.31)	$0.91 ($1.10)	$0.51 ($0.77)
Tax Revenue	$0.48 ($0.17)	$1.60 ($0.75)	$3.82 ($1.75)
Net Government Cost	$1.75 ($1.35)	**-$0.68** ($1.63)	**-$3.31** ($2.25)

Source: Tyner, Brechbill, and Perkis, Purdue University, August 2010.
Notes: Standard deviations are in parentheses. Negative values are in bold.

Stochastic Comparison of a Fixed versus a Variable Subsidy

Given the baseline implied by the previous three tables, it is only meaningful to compare a variable subsidy with a fixed subsidy when a technology shows the potential of being profitable in the fixed subsidy case. Since only grain ethanol was profitable across all oil price forecasts, it is the sole scenario used to compare the different effects of fixed and variable subsidies. Thus, considering only grain ethanol under volumetric pricing, and regardless of the oil price scenario, the standard deviation and the probability of a loss both decrease under a variable subsidy as compared to a fixed subsidy (**Table 26**) suggesting a clear risk-reducing effect.

In addition to reducing the risk of a loss to firms, it is clear that the government's costs are reduced regardless of the oil price outcome, and often to a considerable level (**Table 27**) under a variable subsidy.

CONCLUSIONS

In conclusion, when correctly designed, variable subsidies could reduce risk to firms by decreasing the standard deviation of the net present value. A variable subsidy should also reduce the cost to the government relative to a fixed subsidy, because the government will only subsidize a firm when the oil price is low enough to keep biofuels from being competitive with oil. Fixed subsidies, however, will be paid out regardless of the oil price and whether biofuels could be viable without the subsidy.

When biofuels are priced and subsidized on the basis of energy content, thermochemical conversion could be the least expensive – based on the assumptions in this study. When volumetric pricing of subsidies is used, the higher energy content of biofuels produced via thermochemical conversion will not be properly valued.

All the numeric conclusions in this section, of course, depend on the cost structures, pricing relationships, and other assumptions used in this study. All the numbers come from the literature, but there may be proprietary technologies in the wings with lower costs and/or higher conversion rates.

ACKNOWLEDGMENTS

This technology assessment and report was written by Purdue University, Department of Agricultural Economics, under the leadership of Wallace E. Tyner, together with Sarah Brechbill and David Perkis. The report's authorship rests with Tyner, Brechbill, and Perkis. The work was performed under contract to CRS, and is part of a multi-year CRS project to examine different aspects of U.S. energy policy. This chapter was funded, in part, by a grant from the Joyce Foundation.

End Notes

[1] Wallace E. Tyner, James and Lois Ackerman Professor of Agricultural Economics, Purdue University (West Lafayette, IN), wtyner@purdue.edu, 765-494-0199.

[2] Sarah Brechbill, former graduate student, Department of Agricultural Economics, Purdue University (West Lafayette, IN).

[3] David Perkis, Graduate Research Assistant, Department of Agricultural Economics, Purdue University (West Lafayette, IN), perkis@purdue.edu, 765-494-0593.

[4] Throughout this chapter, the Purdue authors will be referenced as Tyner, Brechbill, and Perkis, Purdue University, August 2010.

[5] J. W. Kram, "Building Blocks to Biofuels Success," *Ethanol Producer Magazine*, December 2008.

[6] E. Petiot, "The Important Role of Enzymes in Cellulosic Ethanol," *Ethanol Producer Magazine*, November 2008.

[7] For more information see CRS Report R40155, *Renewable Fuel Standard (RFS): Overview and Issues*, by Randy Schnepf and Brent D. Yacobucci.

[8] North Central Sun Grant Center, *Composition of Herbaceous Biomass Feedstocks*, Sun Grant Initiative—North Central Center, South Dakota State University (Brookings, SD), June 2007 (hereafter referred to as NC Sun Grant Center (2007)); S. R. Bull, U.S. Department of

Energy, Biofuels Research Program, *Energy Sources, Part A: Recovery, Utilization, and Environmental Effects*, 13(4), 1991, pp. 443-442 (hereafter referred to as Bull (1991)); and N. S. Mosier, "Bioprocess Engineering for Biofuels: Pretreatment and Hydrolysis," Second Generation Biofuels Symposium 2009, Purdue University, West Lafayette, IN, 2009.

[9] For a complete list of crop insurance programs, see "2010 County Crop Programs," Risk Management Agency, USDA, at http://www.rma.usda.gov/data/cropprograms.html.

[10] R. C. Christiansen, "The Cellulosic Ceiling," *Ethanol Producer Magazine*, August 2009.

[11] S. R. Schill, "Cobs to Switchgrass to Gasoline Parity," *Ethanol Producer Magazine*, June 2009.

[12] S. R. Schill, "Oklahoma seeds 1,000 acres of switchgrass," *Biomass Magazine*, July 2008.

[13] L. Gibson and S. Barnhart, "Switchgrass," University Extension, Iowa State University, 2007 (hereafter referred to as Gibson and Barnhart (2007)); and M. Khanna, "Cellulosic Biofuels: Are They Economically Viable and Environmentally Sustainable?" *Choices* 23(3), 2008, pp. 16-2 1 (hereafter referred to as Khanna (2008)).

[14] J. A. Larson and B. C. English. "Risk Management for Energy Investments: Agricultural Policy and Extension Recommendations," *Transition to a Bio Economy: The Role of Extension in Energy*, Farm Foundation Conference (Little Rock, AR), 2009 (hereafter referred to as Larson and English (2009)).

[15] Khanna (2008).

[16] Gibson and Barnhart (2007); D. I. Bransby, *Switchgrass Profile*, Oak Ridge National Laboratory (Oak Ridge, TN), 1999; and C. D. Garland, *Growing and Harvesting Switchgrass for Ethanol Production in Tennessee*, University of Tennessee Biofuels Initiative, Univ. of Tennessee (Knoxville, TN), 2008 (hereafter referred to as Garland (2008)).

[17] CRS Report RS21613, *Conservation Reserve Program: Status and Current Issues*, by Tadlock Cowan.

[18] Larson and English (2009); and M. Duffy, *Estimated Costs for Production, Storage, and Transportation of Switchgrass*, A1-22, Iowa State Univ. (ISU) Extension, ISU (Ames, IA), 2008 (hereafter referred to as Duffy (2008)).

[19] Larson and English (2009).

[20] Brechbill, S.C. and W.E. Tyner, *The Economics of Biomass Collection, Transportation, and Supply to Indiana Cellulosic and Electrical Utility Facilities*, Working Paper 08-03, Department of Agricultural Economics, Purdue University (West Lafayette, IN), April 2008, [hereafter referred to as **Brechbill and Tyner (2008)**].

[21] Ibid.

[22] Schill, S.R., "Collecting Mountains of Stover," *Ethanol Producer Magazine*, 2007.

[23] Bernick, J., "One-Pass Stover Harvest," *AgWeb online Farm Journal*, January 10, 2009, [hereafter referred to as **Bernick (2009)**]; at http://www.agweb.com/article/One-Pass_Stover_Harvest_202067/.

[24] Bernick (2009).

[25] Perry, A., "Cellulosic Ethanol from Corn Stover: Calculating - and Improving - the Bottom Line," *Agricultural Research*, Vol. 56, No. 9, October 2008, pp. 14-15.

[26] State University of New York, College of Environmental Science and Forestry, *Eco WIllow*, v. 1.2, undated, at http://www.esf.edu/willow/default.htm, [hereafter referred to as **State Univ. of New York, (undated)**].

[27] Volk, T.A., T. Verwijst, P.J. Tharakan, L.P. Abrahamson, and E.H. White, "Growing fuel: a sustainability assessment of willow biomass crops," *Frontiers in Ecology and the Environment* 2(8), 2004, p. 411-418.

[28] For examples see CRS Report RL34042, *Provisions Supporting Ecosystem Services Markets in U.S. Farm Bill Legislation*, by Renée Johnson

[29] Taylor (2009).

[30] Hubbard, W., L. Biles, C. Mayfield, S. Ashton (Eds.), *Sustainable Forestry for Bioenergy and Bio-based Products: Trainers Curriculum Notebook*, Southern Forest Research Partnership, Inc, (Athens, GA), September 2007.

[31] CRS Report R40529, *Biomass: Comparison of Definitions in Legislation Through the 111th Congress*, by Kelsi Bracmort and Ross W. Gorte.

[32] U.S. Congress, House Committee on Energy and Commerce, H.R. 2454, *The American Clean Energy and Security Act of 2009*, 111th Congress, 1st sess., passed by the House on June 26, 2009.

[33] Pelkki, M.H., "Technological Trends and Production Costs for Forestry Biomass," in *Transition to a Bio Economy: The Role of Extension in Energy*, Farm Fdn. Conf. (Little Rock, AR), 2009, [hereafter referred to as **Pelkki (2009)**].

[34] Ibid.

[35] Perlack, R.D., L. Wright, A. Turnhollow, R. Graham, B. Stokes, D. Erbach, *Biomass as Feedstock for a Bioenergy and Bioproducts Industry: The Technical Feasibility of a Billion-Ton Annual Supply*, Oak Ridge National Laboratory (Oak Ridge, TN), April 2005, [hereafter referred to as **Perlack et al, *The Billion-Ton Study* (2005)**].

[36] Pelkki, (2009); and Perlack et al, *The Billion-Ton Study* (2005).

[37] National Academies: National Academy of Engineering, National Academy of Sciences, and National Research Council; America's Energy Future Panel on Alternative Liquid Transportation Fuels, *Liquid Transportation Fuels from Coal and Biomass: Technological Status, Costs, and Environmental Impacts*, Washington, DC., 2009, [hereafter referred to as **NAS, *Liquid Transportation Fuels from Coal to Biomass* (2009)**].

[38] U.S. EPA (Washington, D.C.), *Renewable Fuel Standard Program (RFS2) Regulatory Impact Analysis*, EPA-420-R10-006, Assessment and Standards Division, Office of Transportation and Air Quality, 2010.

[39] Gan, J. and C.T. Smith, "Availability of logging residues and potential for electricity production and carbon displacement in the USA," *Biomass and Bioenergy*, 30(12), 2006, pp. 1011-1020.

[40] Tyner, W.E., F. Taheripour, and Y. Han., *Preliminary Analysis of Land Use Impacts of Cellulosic Biofuels*, Argonne National Laboratory and the California Energy Commission, 2009.

[41] Personal communication by Prof. Tyner, Purdue University, with Matt Caswell, BP, undated.

[42] Brechbill and Tyner (2008); Perlack, R.D. and A.F. Turhollow, *Assessment of Options for the Collection, Handling, and Transport of Corn Stover*, ORNL/TM-2002/44, Oak Ridge National Laboratory (Oak Ridge, TN), 2002, [hereafter referred to as **Perlack and Turhollow (2002)**]; Petrolia, D.R., "The economics of harvesting and transporting corn stover for conversion to fuel ethanol: A case study for Minnesota," *Biomass and Bioenergy* 32(7), 2008, p. 603-612, [hereafter referred to as **Petrolia (2008)**]; and Schechinger, T.M. and J. Hettenhaus, *Corn Stover Harvesting: Grower, Custom Operator, and Processor Issues and Answers: Report on Corn Stover Experiences in Iowa and Wisconsin for the 199 7-98 and 1998-99 Crop Years*, ORNL/SUB-04-4500008274-01, Oak Ridge National Laboratory (Oak Ridge, TN), 1999, [hereafter referred to as **Schechinger and Hettenhaus (1999)**].

[43] Petrolia, D.R., "Economics of Crop Residues: Corn Stover," *Transition to a Bio Economy: The Role of Extension in Energy*, Farm Foundation Conference (Little Rock, AR), 2009, [hereafter referred to as **Petrolia (2009)**].

[44] Sokhansanj, S., A. Turhollow, and E. Wilkerson, *Development of the Integrated Biomass Supply Analysis and Logistics Model (IBSAL)*, Oak Ridge National Laboratory (Oak Ridge, TN), 2008, [hereafter referred to as **Sokhansani et al, (2008)**].

[45] Pordesimo, L.O., W.C. Edens, and S. Sokhansanj, "Distribution of above-ground biomass in corn stover," *Biomass and Bioenergy*, 26(4), 2004, pp. 337-343.

[46] Epplin, F., "Alternative Energy and Agriculture: Perspectives on Cellulosic Feedstock and Cellulosic Biorefineries," *Southern Association of Agricultural Sciences* (Altanta, GA), 2009, [hereafter referred to as **Epplin (2009)**].
[47] Petrolia (2009).
[48] Pelkki (2009).
[49] Eidman et al (2009).
[50] *Bioenergy Program*, U.S. Department of Energy, Idaho National Laboratory (INL; Idaho Falls, ID), at https://inlportal.inl.gov/portal/server.pt/community
[51] Hess., J.R., K.L. Kenney, C.T. Wright, R. Perlack, and A. Turhollow, *Corn Stover Availability for Biomass Conversion: Situation Analysis*, INL/JOU-09-15666, Idaho National Laboratory (Idaho Falls, ID), 2009.
[52] Hess, J. R., C.T. Wright, K.L. Kenney, and E.M. Searcy, *Uniform-Format Bioenergy Feedstock Supply System Design Report Series: Commodity-Scale Production Of An Infrastructure-Compatible Bulk Solid From Herbaceous Lignocellulosic Biomass*, "Volume A: 'Uniform-Format' Vision and Conventional-Bale Supply System," Table-2.2, NL/EXT-09-17527DRAFT, Idaho National Laboratory (Idaho Falls, ID), April 2009, p. 26; [hereafter referred to as **Hess et al (2009)**].
[53] NAS, *Liquid Transportation Fuels from Coal to Biomass* (2009).
[54] Rismiller, C.W. and W.E. Tyner, "Transportation Infrastructure Implications of Development of a Cellulosic Biofuels Industry for Indiana," *Journal of the Transportation Research Forum* 49(1), 2010, pp. 95-112.
[55] Petrolia (2009).
[56] Ibid.
[57] Schill, S.R., "Collecting Mountains of Stover," *Ethanol Producer Magazine*, 2007.
[58] Petrolia (2009); and Cundiff, J. "Biomass Logistics in the Southeast," *Transition to a Bio Economy: The Role of Extension in Energy*, Farm Foundation Conference (Little Rock, AR), 2009, [hereafter referred to as **Cundiff (2009)**].
[59] Cundiff (2009).
[60] Schill, S.R., "Miscanthus versus Switchgrass," *Ethanol Producer Magazine*, 2007.
[61] Cundiff (2009).
[62] Petrolia (2009).
[63] Ibid.
[64] Cundiff (2009).
[65] English, B.C. and D.F. Mooney, "Economics of the Switchgrass Supply Chain: Enterprise Budgets and Production Cost Analyses," *Transition to a Bio Economy: The Role of Extension in Energy*, Farm Foundation Conference (Little Rock, AR), 2009, [hereafter referred to as **English and Mooney (2009)**]; and Wang, C., J. Larson, B. English, and K. Jensen., "Cost Analysis of Alternative Harvest, Storage and Transportation Methods for Delivering Switchgrass to a Biorefinery from the Farmers' Perspective," selected paper, *Southern Association of Agricultural Sciences Annual Meeting* (Altanta, GA), 2009.
[66] Wang et al (2009).
[67] Kram, J.W., "Search of Biomass Storage Solutions," *Ethanol Producer Magazine*, 2008, pp. 86-9 1.
[68] Petrolia (2008), Schechinger and Hettenhaus (1999), Glassner et al (1998), Atchinson and Hettenhaus (2003), Khanna et al (2008), and Tiffany et al (2006).
[69] Keoleian, G.A. and T.A. Volk, "Renewable Energy from Willow Biomass Crops: Life Cycle Energy, Environmental and Economic Performance," *Critical Reviews in Plant Sciences* 24, 2005, pp. 385-406.
[70] Badger, P.C., *Processing Cost Analysis for Biomass Feedstocks*, ORNL/TM-2002/199, Oak Ridge National Laboratory (Oak Ridge, TN), October 2002.
[71] Solomon, B.D., J.R. Barnes, and K.E. Halvorsen, "Grain and cellulosic ethanol: History, economics, and energy policy," *Biomass and Bioenergy* 31, 2007, pp. 416-425.

[72] Sims, R., M. Taylor, J. Saddler, and W. Mabee, *From 1st to 2nd Generation Biofuel Technologies: An overview of current industry and RD&D activities*, Organization for Economic Co-Operation and Development and International Energy Agency, © OECD/IEA, November 2008, [hereafter referred to as **Sims et al, OECD/IEA (2008)**].

[73] Rismiller, C.W. and W.E. Tyner, *Cellulosic Biofuels Analysis: Economic Analysis of Alternative Technologies*, Working Paper #09-06, Department of Agricultural Economics, Purdue University, 2009, [hereafter referred to as **Rismiller and Tyner (2009)**].

[74] Sedlak, M. and N.W.Y. Ho, "Production of Ethanol from Cellulosic Biomass Hydrolysate Using Genetically Engineered Saccharomyces Yeast Capable of Co-fermenting Gglucose and Xlyose," *Applicatied Biochemstry and Biotechnology*, Vol. 113-116, 2004, pp. 403-416.

[75] Chapple, C., M. Ladisch, and R. Meilan, "Loosening lignin's grip on biofuel production," *Nature Biotechnology* 25(7), 2007, pp. 746-748.

[76] Hess, J.R., C.T. Wright, and K.L. Kenney, "Cellulosic biomass feedstocks and logistics for ethanol production," *Biofuels, Bioproducts, and Biorefining* 1, 2007, pp. 181-190; and Huber, G.W. and B.E. Dale, "Grassoline: Biofuels beyond Corn," *Scientific American*, July 2009, pp. 52-59, [hereafter referred to as **Huber and Dale (2009)**].

[77] Huber and Dale (2009).

[78] Ibid.

[79] Sims et al, OECD/IEA (2008).

[80] Foust et al (2009)

[81] Ibid.

[82] U.S. EPA (2009), Table 1.5-33, p.165.

[83] Ibid.

[84] Skauge, M., et al., "Proposed Ethanol Plant List 2009: United States and Canada," *Ethanol Producer Magazine*, April 2009.

[85] Christiansen, R.C., "The Cellulosic Ceiling," *Ethanol Producer Magazine*, August, 2009.

[86] U.S. EPA (2009), Table 1.5-36 and Table 1.5-47, pp. 172-173.

[87] Ebert, J., "Cellulosic Ethanol Path is Paved with Various Technologies," *Ethanol Producer Magazine*, July 2008.

[88] DTN *Ethanol Blog*, "Mascoma's Cellulosic Ethanol Plan Likely Delayed," May 25, 2010; and "Iogen Suspends U.S. Cellulosic Ethanol Plant Plans," By Katie Fehrenbacher, gigaom.com, Jun. 4, 2008.

[89] BP and DuPont, "Biobutanol fact sheet," BP-Dupont biofuels fact sheet, June 2006.

[90] BP and DuPont, "Butamax Advanced Biofuels LLC," BP-Dupont biofuels fact sheet, undated.

[91] Agrawal, R., N. Singh, F. Ribeiro, and W.N. Delgass, "Sustainable fuel for the transportation sector," Proceedings of the National Academy of Sciences (PNAS), 104(12), March 2007, pp. 4828-4833.

[92] CRS Report R41282, Agriculture-Based Biofuels: Overview and Emerging Issues, by Randy Schnepf.

[93] In addition to the blending tax credit, the U.S. biofuels industry is supported by an import tariff on foreign-produced ethanol (started in 1980), a mandated blending-use requirement (started in 2006) that expands rapidly to 36 billion gallons by 2022, and several loan, loan guarantee, and grant programs designed to facilitate growth in production and use of biofuels. For more information see CRS Report R401 10, *Biofuels Incentives: A Summary of Federal Programs*, by Brent D. Yacobucci

[94] Tyner, W.E., "The US Ethanol and Biofuels Boom: Its Origins, Current Status, and Future Prospects," *BioScience* 58(7), 2008, pp. 646-653, [hereafter referred to as **Tyner, *Biofuels Origins and Prospects*, (2008)**].

[95] Shapouri, H., and P. Gallagher, *USDA's 2002 Ethanol Cost-of-Production Survey*, AER 841, Office of the Chief Economist, USDA, July 2005.

[96] CRS Report R40155, *Renewable Fuel Standard (RFS): Overview and Issues*, by Randy Schnepf and Brent D. Yacobucci.

[97] Tyner, W. and D. Viteri, "Implications of Blending Limits on the US Ethanol and Biofuels Markets," *Biofuels* 1(2), 2010. pp. 251-253, [hereafter referred to as **Tyner and Viteri (2010)**]; and Tyner, W.E., F. Taheripour, and D. Perkis, "Compariosn of Fixed versus Variable Biofuels Incentives," *Energy Policy*, 38(10), October 2010, pp. 5530-5540, [hereafter referred to as **Tyner, Taheripour, and Perkis (2010)**].

[98] Tyner, W.E., F. Dooley, C. Hurt, and J. Quear, *Ethanol Pricing Issues for 2008, Industrial Fuels and Power*, February 2008, pp. 50-57; and Tyner and Viteri, (2010).

[99] Tyner, W.E., F. Dooley, and D. Viteri, "Effects of Biofuel Mandates in a Context of Ethanol Demand Constraints, Cellulosic Biofuel Costs, and Compliance Mechanisms," *American Journal of Agricultural Economics* 92(5), 2010, [hereafter referred to as **Tyner, Dooley, and Viteri (2010)**].

[100] Renewable Fuel Association, "E85," at http://www.ethanolrfa.org/pages/e-85.

[101] Tyner, Dooley, and Viteri (2010).

[102] DOE, EIA, *Alternatives to Traditional Transportation Fuels 2008*, Table C1. Estimated Consumption of Vehicle Fuels in the United States, by Fuel Type, 2004 – 2008.

[103] CRS Report R40445, *Intermediate-Level Blends of Ethanol in Gasoline, and the Ethanol "Blend Wall"*, by Brent D. Yacobucci.

[104] Ibid.

[105] EPA, "EPA Grants E15 Partial Waiver Decision and Fuel Pump Labeling Proposal," EPA420-F-10-054, October, 2010. at *http://www.epa.gov/otaq/regs/fuels/additive/e15/420f10054.htm.*

[106] Tyner and Viteri (2010), and Tyner, Dooley, and Viteri (2010).

[107] Tyner and Taheripour (2008).

[108] Tyner, Taheripour, and Perkis (2010).

[109] Section 202(c) of EISA, 2007 (P.L. 110-140), as codified in 42 U.S.C. 7545(o)(4).

[110] CRS Report R40155, *Renewable Fuel Standard (RFS): Overview and Issues*, by Randy Schnepf and Brent D. Yacobucci.

[111] U.S. Energy Information Administration, *Updated Annual Energy Outlook 2009*, Department of Energy, September 21, 2009; at [www.eia.doe.gov].

[112] Tyner, W.E., "The Integration of Energy and Agricultural Markets," *International Association of Agricultural Economics Annual Meeting 2009*, (Beijing, China).

[113] CRS Report R401 10, *Biofuels Incentives: A Summary of Federal Programs*, by Brent D. Yacobucci.

[114] CRS Report R41296, *Biomass Crop Assistance Program (BCAP): Status and Issues*, by Megan Stubbs.

[115] Austin, A., "USDA's List of BCAP Qualifiers Grows," *Biomass Magazine*, September2 009.

[116] Grooms, L., "Fiber Farming Comes of Age," *Corn and Soybean Digest*, August2009.

[117] CRS Report R41296, *Biomass Crop Assistance Program (BCAP): Status and Issues*, by Megan Stubbs.

[118] Tyner, *Biofuels Origins and Prospects*, (2008).

[119] CRS Report R401 10, *Biofuels Incentives: A Summary of Federal Programs*, by Brent D. Yacobucci.

[120] CRS Report R41282, *Agriculture-Based Biofuels: Overview and Emerging Issues*, by Randy Schnepf.

[121] CRS Report R401 10, *Biofuels Incentives: A Summary of Federal Programs*, by Brent D. Yacobucci.

[122] CRS Report R41282, *Agriculture-Based Biofuels: Overview and Emerging Issues*, by Randy Schnepf

[123] Tyner, Taheripour, and Perkis (2010); Tyner, W.E. and J. Quear, "Comparison of A Fixed and Variable Corn Ethanol Subsidy," *Choices* 2 1(3), 2006, pp. 199-202; Tyner, W.E. and F. Taheripour, "Renewable Energy Policy Alternatives for the Future," *American Journal of Agricultural Economics*, 89(5), 2007, pp. 1303-1310; and Tyner, W.E. and F. Taheripour,

"Policy Options for Integrated Energy and Agricultural Markets," *Review of Agricultural Economics*, 30(3), 2008, pp. 387-396.

[124] Tyner, Taheripour, and Perkis (2010).

[125] Ibid.

[126] Tyner, Dooley, and Viteri, (2010).

[127] The initial spreadsheet model (done in EXCEL) first appeared in Rismiller and Tyner (2009), but has been modified by D. Perkis of Purdue University for use in this chapter.

[128] This forecasting methodology is outlined in Tyner, Taheripour, and Perkis (2010).

[129] U. S. Energy Information Administration, *Annual Energy Outlook 2010 Early Release Overview*, Dept of Energy (Washington, DC.), 2010.

[130] Tyner, Taheripour, and Perkis (2010).

[131] Tyner, W.E., "The Integration of Energy and Agricultural Markets," *Agricultural Economics*, 41(6), 2010.

[132] From a working model by two of this chapter's authors, Perkis and Tyner.

[133] Tyner, Taheripour, and Perkis (2010).

In: Cellulosic Biofuels: Feedstock, Technology... ISBN: 978-1-61209-509-7
Editors: Robert M. Burkin et al.

Chapter 2

CELLULOSIC BIOFUELS: ANALYSIS OF POLICY ISSUES FOR CONGRESS

Kelsi Bracmort, Randy Schnepf, Megan Stubbs and Brent D. Yacobucci

SUMMARY

Cellulosic biofuels are produced from cellulose (fibrous material) derived from renewable biomass. They are thought by many to hold the key to increased benefits from renewable biofuels because they are made from potentially low-cost, diverse, non-food feedstocks. Cellulosic biofuels could also potentially decrease the fossil energy required to produce ethanol, resulting in lower greenhouse gas emissions.

Cellulosic biofuels are produced on a very small scale at this time—significant hurdles must be overcome before commercial-scale production can occur. The renewable fuels standard (RFS), a major federal incentive, mandated the use of 100 million gallons per year (mgpy) of cellulosic biofuels in 2010. After 2015, most of the increase in the RFS is intended to come from cellulosic biofuels, and by 2022, the mandate for cellulosic biofuels will be 16 billion gallons. Whether these targets can be met is uncertain, and on March 26, 2010, the Environmental Protection Agency issued a final rule that lowers the 2010 cellulosic biofuel mandate to 6.5 million gallons. Research is ongoing, and the cellulosic biofuels industry may be on the verge of rapid expansion and technical breakthroughs. However, at this time, only a few small refineries are scheduled to begin production in 2010, with an additional nine expected to commence production by 2013 for a total output of 389 mgpy, compared with an RFS requirement of 500 mgpy in 2012 (a year earlier).

The federal government, recognizing the risk inherent in commercializing this new technology, has provided loan guarantees, grants, and tax credits in an effort to make the industry competitive by 2012. In particular, the Food, Conservation, and Energy Act of 2008 (the 2008 farm bill, P.L. 110-246) supports the nascent cellulosic industry through authorized research programs, grants, and loans exceeding $1 billion. The enacted farm bill also contains a production tax credit of $1.01 per gallon for ethanol produced from cellulosic feedstocks. Private investment, in many cases by oil companies, also plays a major role in cellulosic biofuels research and development.

Three challenges must be overcome if the RFS is to be met. First, cellulosic feedstocks must be available in large volumes when needed by refineries. Second, the cost of converting cellulose to ethanol or other biofuels must be reduced to a level to make it competitive with gasoline and corn-starch ethanol. Third, the marketing, distribution, and vehicle infrastructure must absorb the increasing volumes of renewable fuel, including cellulosic fuel mandated by the RFS.

Congress will continue to face questions about the appropriate level of intervention in the cellulosic industry as it debates both the risks in trying to pick the winning technology and the benefits of providing start-up incentives. The current tax credit for cellulosic biofuels is set to expire in 2012, but its extension may be considered during the 111th Congress. Congress may continue to debate the role of biofuels in food price inflation and whether cellulosic biofuels can alleviate its impacts. Recent congressional action on cellulosic biofuels has focused on the definition of renewable biomass eligible for the RFS, which is considered by some to be overly restrictive. To this end, legislation has been introduced to expand the definition of renewable biomass eligible under the RFS.

INTRODUCTION

Cellulosic biofuels are produced from cellulose[1] derived from renewable biomass feedstocks such as corn stover (plant matter left in the field after harvest), switchgrass, wood chips, and other plant or waste matter. Current production consists of a few small-scale pilot projects—and significant hurdles must be overcome before industrial-scale production can occur.

Ethanol produced from corn starch and biodiesel produced from vegetable oil (primarily soybean oil) are currently the primary U.S. biofuels.[2] High oil and gasoline prices, environmental concerns, rural development, and national energy security have driven interest in domestic biofuels for many years. However, the volume of fuel that can be produced using traditional row crops such as corn and soybeans without causing major market disruptions is limited; to fulfill stated goals, biofuels must also come from other sources that do not compete for the same land used by major food crops. Proponents see cellulosic biofuels as a potential solution to these challenges and support government incentives and private investment to hasten efforts toward commercial production. Some federal incentives—grants, loans, tax credits, and direct government research—attempt to push cellulosic biofuels technology to the marketplace. Demand-pull mechanisms such as the renewable fuel standard (RFS) mandate the use of biofuel blends—creating an incentive for the development of a new technology to enter the marketplace.

In contrast, petroleum industry critics of biofuel incentives argue that technological advances such as seismography, drilling, and extraction continue to expand the fossil-fuel resource base, which has traditionally been cheaper and more accessible than biofuel supplies. Other critics argue that current biofuel production strategies can only be economically competitive with existing fossil fuels in the absence of subsidies if significant improvements are made to existing technologies or new technologies are developed. Until such technological breakthroughs are achieved, critics contend that the subsidies distort energy markets and divert research funds from the development of other renewable energy sources not dependent on internal combustion technology, such as wind, solar, or geothermal, which offer potentially cleaner, more bountiful alternatives. Still others debate the rationale behind policies that promote biofuels for energy security, questioning whether the United States could ever produce and manage sufficient feedstocks of starches, sugars, vegetable oils, or even cellulose to permit biofuel production to meaningfully offset petroleum imports. Finally, there are those who argue that the focus on development of alternative energy sources undermines efforts to score energy savings through lower consumption.

The Renewable Fuel Standard: A Mandatory Usage Mandate

Principal among the cellulosic biofuels goals to be met is a biofuels usage mandate—the renewable fuel standard (RFS) as expanded by the Energy Independence and Security Act of 2007 (EISA, P.L. 110-140, Section 202)—that includes a specific carve-out for cellulosic biofuels.[3] The RFS is a demand-pull mechanism that requires a minimum usage of biofuels in the nation's fuel supply. This mandate can be met using a wide array of technologies and fuels.

Although most of the RFS is expected to be met using corn ethanol initially, over time the share of advanced (non-corn-starch derived) biofuels in meeting the mandate increases. The RFS specifies three non-corn-starch carve-outs: cellulosic biofuels, biomass-based diesel fuel, and other (or unspecified), which could potentially be met by imports of sugarcane-based ethanol. The RFS mandate for cellulosic biofuels in the EISA begins at 100 million gallons per year in 2010 and rises to 16 billion gallons per year in 2022 (**Figure 1**).[4] This mandate represents a prodigious challenge to the biofuels industry in light of the fact that no commercial production of cellulosic biofuels yet exists in the United States. Indeed, on March 26, 2010, the U.S. Environmental Protection Agency (EPA) issued a final rule for implementation of the RFS that sets a new, lower cellulosic biofuel mandate of 6.5 million gallons for 2010.

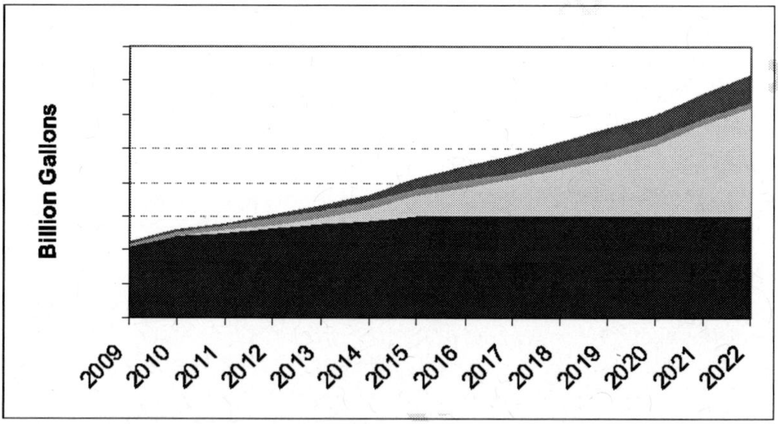

Source: EISA, (P.L. 110-140, Section 202).

Notes: Corn-starch ethanol volume is a cap, whereas other categories are floors. Biodiesel includes any type of biomass-based diesel substitute.

Figure 1. Renewable Fuel Standard Under EISA as of November 2009

The RFS also mandates maximum lifecycle greenhouse gas emissions for each type of biofuel. Lifecycle greenhouse gas emissions encompass emissions[5] at all levels of production, from the field to retail sale, including emissions resulting from land use changes, that is, the clearing of forests for cropland due to increased energy crop production elsewhere. Under the law, GHG emissions for cellulosic biofuels qualifying for the RFS are limited to 60% of the GHG emissions from extracting, refining, distributing, and consuming gasoline.[6]

Challenges Facing the Industry

Cellulosic biofuels have potential, but there are significant hurdles to overcome before competitiveness is reached. In his 2007 State of the Union Address, President Bush announced the "Twenty in Ten" initiative, calling for the rapid expansion of renewable biofuels production as a major part of an effort to reduce U.S. gasoline use by 20% through biofuels and conservation. This goal was given substance in December 2007, when Congress passed EISA, mandating the RFS for the use of specific volumes of renewable biofuels through 2022 and setting a goal of commercial-scale cellulosic biofuels production by 2012.

This chapter provides background on the current effort to develop industrial-scale, competitive technology to produce biofuels from cellulosic feedstocks. It outlines the three major challenges faced in the context of the RFS: (1) feedstock supply, (2) extraction of fuel from cellulose, and (3) biofuel distribution and marketing issues. It then examines the current role of government (in cooperation with private industry) in developing that technology. Finally, the report reviews the role of Congress with respect to the emerging cellulosic biofuels industry, reviews recent congressional actions affecting the industry, and discusses key questions facing Congress.

Cellulosic Feedstock Supplies

Feedstocks used for cellulosic biofuels are potentially abundant and diverse. Initially it was thought that a major advantage of cellulosic biofuels over corn-starch ethanol was that they could be derived from potentially inexpensive feedstocks that could be produced on marginal land.[7] Corn, on the other hand, is a resource-intensive crop that requires significant use of chemicals, fertilizers, and water, and is generally grown on prime farmland. However, field research now suggests that establishment costs, as well as collection, storage, and transportation costs, associated with the production of bulky biomass crops are likely to be more challenging than originally thought.[8]

Cellulose, combined with hemicellulose and lignin, provides structural rigidity to plants and is also present in plant-derived products such as paper and cardboard. Feedstocks high in cellulose come from agricultural, forest, and even urban sources (see **Table 1**). Agricultural sources include crop residues and biomass crops such as switchgrass; forest sources include tree plantations, natural forests, and cuttings from forest management operations. Municipal solid waste, usually from landfills, is the primary urban source of renewable biomass.

Table 1. Potential Cellulosic Feedstock Sources

Source	Feedstock
Agricultural crop residues	Crop residues—stover, straw, etc.
Agricultural commercial crops	Perennial prairie grasses
Forest woody biomass	Logging residues from conventional harvest operations and forest management and land clearing operations
	Removal of excess biomass from timberlands and other forest lands
	Fuelwood from forest lands
	Perennial woody crops
Agricultural or forest processing by-products	Food/feed processing residues
	Pulping (black) liquor from paper mills
	Primary and secondary wood processing mill residues
Urban	Municipal solid Waste
	Packaging wastes and construction debris

Source: CRS.

Cellulosic feedstocks may have some environmental drawbacks. Some crops suggested for biomass are invasive species when planted in non-native environments. Municipal solid wastes may likely require extensive sorting to segregate usable material and may also contain hazardous material that is expensive to remove. In general, calculation of the estimated cost of biofuels production does not reflect environmental or related impacts, but such impacts are relevant to overall consideration of biofuels issues.

Biomass feedstocks are bulky and difficult to handle, presenting the industry with a major challenge. Whether feedstocks are obtained from agriculture or forests, specialized machinery would need to be developed to harvest and handle large volumes of bulky biomass. For instance, harvesting corn for both grain and stover would be more efficient with a one-pass machine capable of simultaneously segregating and processing both—a combination forage and grain harvester. Currently, machines such as these are being developed to handle biomass crops, but few are commercially available.[9] Storage facilities capable of keeping immense volumes of cellulosic material in optimal conditions must also be developed, if an industry is to grow.

Crop Residues

Crop residues are by-products of production processes (such as producing grain), and so their production costs are minimal. Corn stover[10] and rice and wheat straw are abundant agricultural residues with biomass potential.[11] Among residues, corn stover has attracted the most attention for biofuels production. However, an important indirect cost associated with using crop residue as a biomass feedstock is a potential loss of soil fertility. When harvesting stover, sufficient crop residue must be left in place to prevent erosion and maintain soil fertility. Research suggests that, under the right soil conditions, up to 60% of some residuals can be removed without detrimental soil nutrition or erosion effects. Results from early trials suggest the potential ethanol yield from corn stover (not including the grain harvested, which could be used for feed or fuel) is approximately 180 gallons of ethanol per acre. This compares with roughly 425 gallons of cornstarch ethanol[12] (from grain) and 662 gallons per acre of sugar cane (in Brazil), when grown as dedicated energy crops.[13]

Prairie Grasses

Perennial prairie grasses include native species, which were common before the spread of agriculture, and non-indigenous species, some of which are now quite common. Switchgrass is a native perennial grass that once covered American prairies and is a potential source of biomass. Its high density and native immunity to diseases and pests have caused many to focus on its use as a cellulosic feedstock. According to research at the University of Tennessee, the 10-foot tall grass, if harvested after frost, will produce for 10 to 20 years. However, like other perennials, switchgrass takes some time to establish—according to field trials, in the first year of production, yields are estimated at 30% (two tons per acre) of the full yield potential. In the second year, yield is about 70% (five tons per acre), and in the third year yields reach full potential at seven tons per acre,[14] the equivalent of 500[15] to 1,000 gallons of ethanol.[16]

Miscanthus is another fast-growing perennial grass. Originally from Asia, it is now common in the United States. Miscanthus produces green leaves early in the planting season and retains them through early fall, maximizing the production of biomass.[17] Like switchgrass, it grows on marginal lands with minimal inputs. Research in Illinois shows miscanthus can produce 2½ times the volume of ethanol (about 1,100 gallons) per acre as corn—under proper conditions.[18]

At South Dakota State University, field trials with mixtures of native grasses produced biomass yields slightly lower than switchgrass monocultures, but suggest that such mixtures result in better soil health, improved water quality, and better wildlife habitat.[19] Similar research at the University of Minnesota with mixtures of 18 native prairie species resulted in biomass yields three times greater than switchgrass.[20]

Forest Sources of Biomass

Forest resources for biomass include naturally occurring trees, residues from logging and other removals, and residue from fire prevention treatments. Extracting and processing forest biomass can be expensive because of poor accessibility, transportation, and labor availability. More efficient and specialized equipment than currently exists is needed for forest residual recovery to become cost effective.[21]

Commercial tree plantations (perennial woody crops) are another source of woody biomass. Compared to prairie grasses, perennial woody crops such as hybrid poplar, willow, and eucalyptus trees, are relatively slow to mature and require harvesting at long intervals (2-4 year intervals for willow or 8-15 years for poplar). Using specialized equipment, harvesting usually occurs in the winter, when trees are converted to chips on site and then transported to the refinery for processing. Some trees, such as willow, re-sprout after cutting and can be harvested again after a few years.[22] An acre of woody biomass (i.e., hybrid poplar) yields an estimated 700 gallons of biofuel on an annual basis.[23]

Secondary and Tertiary Feedstocks

Secondary and tertiary feedstocks are derived from manufacturing (secondary) or consumer (tertiary) sources. In many cases their use as feedstocks recovers value from low- or negative- value materials. Food and feed processing residues such as citrus skins are major agricultural residues often suitable as renewable biomass. Residues from wood processing industries such as paper mills or from feed processing are major secondary sources. Tertiary sources include urban wood residues such as construction debris, urban tree trimmings, packaging waste, and municipal solid waste. One ton of dry woody biomass produces approximately 70 gallons of biofuels.[24]

Feedstock Issues

Volumes Required

Ethanol plants are intended to operate 24/7, that is, year-round with only a brief temporary stoppage for maintenance. As a result, accumulating and storing enough feedstock to supply a commercial-scale refinery producing 10-20 million gallons per year (mgpy) would require as much as 700 tons of feedstock a day—nearly the volume of 900 large round bales of grass or hay—or about 240,000 tons annually.[25] In contrast, a (much larger) 100 mgpy corn ethanol plant requires about 2,500 tons of corn per day, but corn is much denser and easier to handle than most renewable biomass sources.[26] The U.S. Department of Energy (DOE) is currently focusing research efforts on harvest and collection, preprocessing, storage and queuing, handling, and transportation of feedstocks.[27] These are major challenges facing an emerging biofuels industry due to the sheer bulk of the biomass and divergent growth cycles of different biomass crops.

Pelletizing and other methods for compressing feedstocks reduce transportation costs but increase processing costs. According to a Purdue University study, the total per ton costs for transporting biomass 30 miles range from $39 to $46 for corn stover and $57 to $63 for switchgrass—compared with roughly $10 for corn.[28] The USDA-DOE goal is to reduce the total feedstock cost at the plant (production, harvest, transport, and storage) from $60 per ton (the 2007 level) to $46 per ton in 2012.[29]

A 2005 USDA-DOE study undertaken by the Oak Ridge National Laboratory estimates that just over 1.3 billion tons of biomass (**Figure 2**) could be available annually in the United States for all forms of bioenergy production (including electricity and power from biomass, and fuels from cellulose).[30] If processed into biofuel, this 1.3 billion tons of biomass could replace 30% of U.S. transportation fuel consumption at 2004 levels, according to USDA. However, this estimate has been heavily criticized for several reasons, including the claim that it ignores the costs and difficulties likely to be associated with harvesting or collecting woody biomass and urban waste, as well as that it uses optimistic yield growth assumptions to achieve its biomass tonnages. The USDA estimate also predates the definition of renewable biomass eligible for the RFS. Current provisions restrict the use of woody biomass to trees grown in plantations or pre-commercial thinnings from non-federal lands, while USDA's study included woody biomass from federal and private forests as well as commercial forests. As a result, the potential volume of biomass available for conversion is substantially less than the USDA-DOE estimate of 1.3 billion tons.

Impacts on Food Supplies

Compared with corn, cellulosic feedstocks are thought to have smaller impacts on food supplies.[31] By refining corn into ethanol, food markets are indirectly affected via cattle and dairy feed markets. In contrast, cellulosic feedstocks are non-food commodities and thus do not reduce food output unless they displace food crops on cropland.[32] However, many cellulosic feedstocks do not need prime farmland. Waste streams such as municipal solid waste, most crop residues, wood pulp residues, and forest residues are potential sources of biomass that have no impact on food crop acreage.[33] Corn stover, removed in appropriate quantities, could also be refined into ethanol without affecting food supplies. Feedstocks such as switchgrass and fast-growing trees appear to do well in marginal conditions and would likely have a minimal impact on food supplies, particularly in the case of woody biomass feedstocks from forested areas not suitable for crops.[34]

Establishment Costs and Contracting Arrangements

In the United States, crops are traditionally grown on an annual basis. Thus, contracts, loans, and other arrangements are generally established for a single growing year. Arrangements for producing perennial crops would necessarily reflect their multi-year cycles. Producers, whether they own or rent land, can expect reduced returns while the crop becomes established. Producers renting land would need long-term agreements suitable for multi-year crops. Some suggest a legal framework would have to be developed for multi-year harvests. For example, the University of Tennessee has entered into three-year contracts with producers to ensure switchgrass availability for a pilot refinery scheduled to have started ethanol production in 2009.[35]

EXTRACTING FUEL FROM CELLULOSE: CONVERSION

Breaking down cellulose and converting it into fuel requires complex chemical processing— technology that is now rudimentary and expensive (see **Table 2**). Starches (such as corn) and sugars (such as cane sugars) are easily fermented into alcohol, but cellulose must first be separated from hemicellulose and lignin and then broken down into sugars or starches through enzymatic processes.[36] Alternatively, biomass can be thermochemically converted into synthesis gas (syngas),[37] which can then be used to produce a variety of fuels. Regardless of the pathway, as discussed below, at the present time processing cellulose into fuels is expensive relative to other conventional and alternative fuel options.

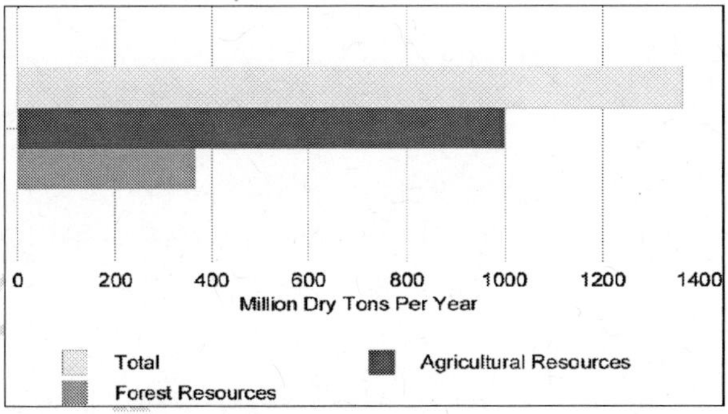

Source: Oak Ridge National Laboratory, 2005.
Note: Total is roughly equivalent to 42 billion gallons of gasoline.

Figure 2. Annual Biomass Resource Potential According to USDA

Table 2. Basic Steps Required to Produce Ethanol

Product	Feedstock	Refining Required after Milling			Cost (per gal.)
Sugar ethanol	Sugar cane		Fermentation into ethanol		$0.30
Corn-starch ethanol	Corn starch	Hydrolysis makes fermentable sugars	Fermentation into ethanol		$0.53
Cellulosic ethanol (biochemical process)	Switchgrass, corn stover, woody bio-mass, municipal solid waste	Pre-treatment makes cellulose accessible	Hydrolysis makes fermentable sugars	Fermentation into ethanol	$1.59
Cellulosic ethanol (therm-chemical process)	Switchgrass, corn stover, woody bio-mass, municipal solid waste	Pre-treatment makes cellulose accessible	Syngas production through therm-chemical processes	Conversion of syngas to products including ethanol	$1.21

Source: National Renewable Energy Laboratory (NREL), Research Advances in Cellulosic Ethanol, costs from the *Economic Feasibility of Ethanol Production From Sugar in the United States*, Hosein Shapouri, USDA, and Michael Salassi, Nelson Fairbanks, LSU Agricultural Center, March 2005.

Production Processes

Three basic methods can be used to convert cellulose into ethanol: (1) acid hydrolysis (dilute or concentrated), (2) enzymatic hydrolysis, and (3) thermochemical gasification and pyrolysis. There are many different variations on these, depending on the enzymes and processes used. Currently all these methods are limited to pilot or demonstration plants, and all comprise the "pre-treatment" phase of ethanol production.

Acid Hydrolysis

Dilute and concentrated acid hydrolysis pre-treatments use sulphuric acid to separate cellulose from lignin and hemicellulose. Dilute acid hydrolysis breaks down cellulose using acid at high temperature and pressure. Only about

50% of the sugar is recovered because harsh conditions and further reactions degrade a portion of the sugar. In addition, the combination of acid, high temperature, and pressure increase the need for more expensive equipment.

On the other hand, concentrated acid hydrolysis occurs at low temperature and pressure and requires less expensive equipment. Although sugar recovery of over 90% is possible, the process is not economical, due to extended processing times and the need to recover large volumes of acid.[38]

Enzymatic Hydrolysis

DOE suggests that enzymatic hydrolysis, a biochemical process that converts cellulose into sugar using cellulase enzymes, offers both processing advantages as well as the greatest potential for cost reductions.[39] However, the cost of cellulase enzymes remains a significant barrier to the conversion of lignocellulosic biomass to fuels and chemicals. Enzyme cost primarily depends on the direct cost of enzyme preparation ($/kg enzyme protein) and the enzyme loading required to achieve the target level of cellulose hydrolysis (gram enzyme protein/gram cellulose). According to DOE, the near-term goal is to reduce the cost of cellulase enzymes from $0.50 to $0.60 per gallon of ethanol to approximately $0.10 per gallon.[40] The National Renewable Energy Laboratory (NREL) of DOE is conducting research to lower enzyme costs by allowing cellulase yeasts and fermenting yeasts to work simultaneously—with significant savings.

The total conversion cost (excluding feedstock cost) for biochemical conversion of corn stover to ethanol is estimated to be about $1.59 per gallon[41]—compared with the USDA-DOE goal of $0.82 per gallon in 2012.[42]

Thermochemical Gasification and Pyrolysis

Thermochemical processes such as gasification and pyrolysis convert lignocellulosic biomass into a gas or liquid intermediate (syngas) suitable for further refining to a wide range of products including ethanol, diesel, methane, or butanol.[43] Recovery rates of up to 50% of the potentially available ethanol have been obtained using synthesis gas-to-ethanol processes. Two-stage processes producing methanol as an intermediate product have reached efficiencies of 80%. However, developing a cost-effective thermochemical process has been difficult.[44] The Fischer-Tropsch (FT) process uses gasification to produce syngas that is then converted into biofuels such as diesel, methane, or butanol. It is possible to produce diesel and other fuels using syngas from coal or natural gas, but biomass-derived syngas is technically challenging because of impurities that must be removed during processing.

The cost of gasification conversion (excluding the cost of feedstock) in 2005 was estimated at $1.21 per gallon (2007 dollars).[45] The USDA-DOE goal for 2012 is $0.82 cents per gallon.

DISTRIBUTION AND ABSORPTION CONSTRAINTS

Distribution and absorption constraints may hinder the use of cellulosic biofuels even if they are ultimately produced on an industrial scale. In the coming years, greater volumes of advanced biofuels (i.e., cellulosic or non-corn-starch ethanol, biodiesel, or imported sugar ethanol) would need to be blended into motor fuel to fulfill the rising advanced biofuel mandate.

Distribution Bottlenecks

Distribution issues may hinder the efficient delivery of ethanol to retail outlets. Ethanol, mostly produced in the Midwest, would need to be transported to more populated areas for sale. It cannot currently be shipped in pipelines designed for gasoline because it tends to attract water in gasoline pipelines.

In addition, ethanol must be stored in unique storage tanks and blended prior to delivery to the retail outlet, because it tends to separate if allowed to sit for an extended period after blending. This would require further infrastructure investments.

The current ethanol distribution system is dependent on rail cars, tanker trucks, and barges. Because of competition, options (especially rail cars) are often limited. As non-corn biofuels play a larger role, some infrastructure concerns may be alleviated as production is more widely dispersed across the nation. If biomass-based diesel substitutes are produced in much larger quantities, some of these infrastructure issues may be mitigated.

The Blend Wall[46]

The blend wall refers to the volume of ethanol required if all gasoline used in the United States contained 10% ethanol (E-10)[47]—or roughly 14 billion gallons. However, because of infrastructure issues associated with transporting and storing midwestern ethanol in coastal markets, the effective blend wall is probably about 12 billion gallons per year. U.S. ethanol production is rapidly approaching this level. Once the blend wall is reached, the market will have difficulty absorbing further production increases, even if they are mandated by the RFS. Although greater use of E-85 could absorb additional volume, it is limited by the lack of E-85 infrastructure (including the considerable expense of installing or upgrading tanks and pumps) and the size of the flex-fuel fleet. These concerns could be sidestepped if additional non-ethanol biofuels are introduced into the market, especially "drop-in" fuels that are chemically similar to petroleum fuels and could be blended directly with those fuels.[48]

EPA considered proposals to raise the ethanol blend level for conventional vehicles from E- 10 to E-15 or E-20 after a petition was submitted by Growth Energy in 2009. In October 2010, EPA issued a waiver for fuel to contain up to 15% ethanol (E1 5) for model year 2007 and newer light- duty vehicles.[49] A decision on the use of E1 5 in model year 2001-2006 vehicles will be made after EPA receives the results of additional DOE testing, perhaps as early as November. However, no waiver is being granted for E1 5 use in model year 2000 and older cars and light trucks—or in any motorcycles, heavy-duty vehicles, or non-road engines—because currently there is no testing data to support such a waiver. In addition to the EPA waiver announcement, numerous other changes have to occur before gas stations will begin selling E15, including many approvals by states and significant infrastructure changes (pumps, storage tanks, etc.). As a result, the vehicle limitation to newer models, coupled with infrastructure issues, is likely to limit rapid expansion of blending rates. Along with the waiver, EPA issued a notice of proposed rulemaking (NPRM) to promote the successful introduction of E15 into commerce by ensuring that E15 is used in approved motor vehicles and reducing the potential for the misfueling of E15 into vehicles and engines for which it is not approved.

ECONOMIC AND ENVIRONMENTAL ISSUES

Economic Efficiency

Cellulosic biofuels are generally thought to have favorable economic efficiency potential over corn-starch ethanol primarily because of the low costs of production for feedstocks. However, current NREL estimates of the total cost of producing cellulosic ethanol, including feedstock production, marketing, and conversion, are $2.40 per gallon, more than twice the cost of producing corn ethanol.

A major impediment to the development of a cellulose-based ethanol industry is the state of cellulosic conversion technology (i.e., the process of gasifying cellulose-based feedstocks or converting them into fermentable sugars).[50] DOE's goal of competitiveness in 2012 assumes $1.30 (2007 dollars) per gallon costs for corn stover ethanol based on a feedstock price of $13 per ton. This compares with USDA's estimated cost of producing corn-based ethanol in 2002 of $0.958 per gallon (about $1.07 per gallon in 2007 dollars).[51] In addition, the cost of harvesting, transporting, and storing bulky cellulosic biomass is not well understood and consequently is often undervalued. As a result, even though cellulosic ethanol benefits from the $1.01 production tax credit (discussed below), which is $0.56 per gallon higher than the blender's tax credit of $0.45 per gallon for corn ethanol, it remains at a substantial cost disadvantage compared with corn-starch ethanol.

Energy Balance

The net energy balance (NEB) is a comparison of the ratio of the per-unit energy produced versus the fossil energy used in a biofuel's production process. The use of cellulosic biomass in the production of biofuels yields an improvement in NEB compared with corn ethanol. Corn ethanol's NEB was estimated at 67% by USDA in 2004—67% more energy was available in the ethanol than was contained in the fossil fuel used to produce it. This is at the upper range of estimates for corn ethanol's energy balance. By contrast, estimates of the NEB for cellulosic biomass range from 300%[52] to 900%.[53] As with corn-based ethanol, the NEB varies based on the production process used to grow, harvest, and process feedstocks.

Another factor that favors cellulosic ethanol's energy balance over corn-based ethanol relates to byproducts. Corn-based ethanol's co-products are valued as animal feeds, whereas cellulosic ethanol's co-products, especially lignin, are expected to serve directly as a processing fuel at the plant, substantially increasing energy efficiencies.

Additionally, switchgrass uses less fertilizer than corn, by a factor of two or three,[54] and its perennial growth cycle reduces field passes for planting. Some suggest that ethanol from switchgrass has at least 700% more energy output per gallon than fossil energy input.[55] The same is largely true of woody biomass that, even in plantations, requires minimal fertilizer and infrequent planting operations.

Environment

Ethanol and biodiesel produced from cellulosic feedstocks, such as prairie grasses and fast- growing trees, have the potential to improve the energy and environmental effects of U.S. biofuels. As previously stated, a key potential benefit of cellulosic feedstocks is that they can be grown without the need for chemicals. Reducing or eliminating the need for chemical fertilizers could address one of the largest energy inputs for corn-based ethanol production. Fast-growing trees and woody crops could offer additional environmental benefits of improved soil and water quality, reduced CO_2 emissions, and enhanced biodiversity.[56]

Despite potential environmental benefits, additional concerns about cellulosic feedstocks exist, including concerns that required increases in per-acre yields to obtain economic feasibility could require the use of fertilizers or water resources, and that availability of sufficient feedstock supply is limited and expansion could generate additional land use pressures for expanded production (see "Greenhouse Gas Emissions" discussion, below). In addition to these concerns, some groups say that other potential environmental drawbacks associated with cellulosic fuels should be addressed, such as the potential for soil erosion, increased runoff, the spread of invasive species, and disruption of wildlife habitat.

Greenhouse Gas Emissions

Greenhouse gas emissions differ among types of ethanol because of a number of factors, including the feedstock crop converted into ethanol, the fuel used to power the refinery (fossil or renewable), and the original state of the land on which the feedstock was produced. For instance, if virgin forest land were cleared and planted with switchgrass, higher greenhouse gas emissions would result than if switchgrass were grown on previously cleared cropland, mainly because GHG emissions associated with clearing and plowing the virgin soil would have to be included as part of the production process. Likewise, a cellulosic refinery powered by coal or natural gas would have higher greenhouse gas emissions than one powered by recovered feedstock co- products.

Multi-year harvesting of perennial crops decreases greenhouse gas emissions by minimizing field passes. Prairie grasses and woody crops require reduced inputs compared with corn—and have lower greenhouse gas emissions. Also, because cellulosic feedstocks require less fertilizer for their production, the energy balance benefit of cellulosic ethanol could be significant. A study by the Argonne National Laboratory concluded that with advances in technology, the use of herbaceous[57]-feedstock cellulose-based E-10 could reduce fossil energy consumption per mile by 8%, while herbaceous-feedstock cellulose-based E-85 could reduce fossil energy consumption by roughly 70%.[58]

According to the EPA's Office of Transportation and Air Quality, for every unit of energy measured by British Thermal Units (BTU) of gasoline replaced by cellulosic ethanol, the total lifecycle greenhouse gas emissions (including carbon dioxide, methane, and nitrous oxide) would be reduced by an average of about 90%. In comparison, the reduction from corn ethanol averages 22%.[59]

Private Investment

Private investment is viewed by many to be critical to the development of the cellulosic biofuels industry. However, the aggregate value of required private investment is difficult to determine. Anecdotal evidence suggests the main sources of capital are venture capitalists and petroleum companies—commercial banks have a minor role. Venture capitalists generally have an extended (10-year) perspective, which fits well with nascent technologies and is insulated from shorter- term financial volatility. Petroleum companies, faced with mandatory blending of biofuels with gasoline, have been eager to invest in the cellulosic industry. Numerous partnerships have been formed: British Petroleum (BP) and Verenium announced a partnership in August 2008 to accelerate the commercialization of cellulosic ethanol, with BP investing $90 million in the deal.[60] In another collaboration, Royal Dutch Shell has teamed up with Imogen Corporation to develop cellulosic ethanol processes.[61] Mascoma, a major ethanol producer, raised $30 million to support its investment in cellulosic feedstock conversion with technical support from General Motors and Marathon Oil.[62] A collaboration between Monsanto and Mendel Biotechnology Inc. will focus on the breeding and development of crops for production of cellulosic biofuels.[63]

FEDERAL CELLULOSIC BIOFUELS POLICIES

USDA and DOE are currently engaged in a variety of activities to encourage development and demonstration of cellulosic biofuels technologies. The Energy Independence and Security Act of 2007 (EISA, P.L. 110-140), the Food, Conservation, and Energy Act of 2008 (the 2008 farm bill, P.L. 110-246), and other legislation support research and development of a broad range of cellulosic technologies through USDA and DOE programs. Many of these programs extend the goals of the Energy Policy Act of 2005 (EPAct, P.L. 109-58) and President Bush's *20 in 10* initiative.[64] The Biomass Research and Development Initiative (BRDI) coordinates federal interagency *technology-push* efforts, such as R&D, loans, and grants, under the guidance of the Biomass Research and Development Board. The board was authorized in the Biomass Research and Development Act of 2000 and is co-chaired by USDA and DOE. BRDI plays a major role in R&D for the cellulosic biofuels industry.[65]

In October 2008, then-USDA Secretary Ed Schafer and DOE Secretary Samuel W. Bodman released the National Biofuels Action Plan (NBAP), which provided an outline of the major challenges facing the cellulosic biofuels industry: feedstock production and logistics; conversion science and technology; distribution infrastructure and blending. The plan reflects current federal and industry efforts to develop the cellulosic biofuels industry.[66]

Direct Federal Spending on R&D

Recognizing that cellulosic biofuels can contribute to improving national energy security, reducing greenhouse gas emissions, and boosting rural economic development, discretionary DOE spending on bioenergy R&D (including a major cellulosic component) under the Biomass and Biorefinery Systems Program (Energy Efficiency and Renewable Energy Programs) was $196 million in FY2007.[67] DOE appropriations for this purpose totaled $198 million in FY2008, of which 33% was spent on conversion R&D, 7% on feedstock infrastructure, and 52% on biorefinery development.[68] DOE appropriations for 2009 amounted to $217 million, of which $215 million was directed toward cellulosic biofuels, plus an FY2009 stimulus of $786 million. DOE appropriations for FY20 10 are $220 million.[69] Approximately $3 million in congressionally directed spending for FY2010 will be directed toward cellulosic biofuel initiatives.[70]

USDA discretionary outlays for Bioenergy and Renewable Energy Programs, which funded cellulosic biofuels in part, were nearly $100 million in FY2008, and the FY2009 budget request is $82 million.[71] USDA R&D expenditures were estimated at $39 million in FY2008, and budgeted at $59 million in FY2009. Commercialization outlays (primarily the Bioenergy Program) totaled an estimated $59 million in FY2008, and are budgeted at $18 million in FY2009. These totals are modest in comparison to the $5 to $8 billion in annual federal support for corn ethanol. Over time, as the corn ethanol industry matures, the focus of USDA efforts is likely to increasingly shift to cellulosic biofuels.

Federal-Private Partnerships

Private sector investment received a substantial federal policy boost on February 28, 2007, when the DOE announced the awarding of up to $385 million in mandatory cost-share funding for the construction of six cellulosic ethanol plant projects over a four-year period under Section 932 of the EPAct of 2005 as expanded by EISA of 2007. When fully operational, the six plants combined were expected to produce up to 100 mgpy of cellulosic ethanol. These demonstration-scale biorefinery projects focus on near-term commercial processes. The combined cost-share plus federal funding for the projects represents total planned investment of more than $1.2 billion.

The uncertainties of moving an industry from laboratory to commercial reality were highlighted when two recipients with total grant funding of $113 million dropped out of the program, one because of a substantially higher offer from the Canadian government,[72] and the other after determining that the risks involved outweighed any anticipated benefits.[73]

Renewable Energy Provisions in the 2008 Farm Bill (P.L. 110-246)

Renewable energy policy in the 2008 farm bill (Food, Conservation, and Energy Act of 2008, P.L. 110-246) builds on earlier programs originally authorized in the 2002 farm bill (P.L. 107-171) or the EPAct of 2005 (P.L. 109-58) but provides greater emphasis on cellulosic biofuels.[74] Title IX, the energy title, authorizes or reauthorizes grants, loans, and loan guarantees to foster research on agriculture-based renewable energy, to share development risk, and to promote the adoption of renewable energy systems.[75] Implementation of the farm bill provisions is underway, and regulations for new programs have not been finalized. Funding for the cellulosic component of renewable energy programs is difficult to determine because most provide support to a wide range of biofuels. Title VII, the research title, contains provisions supporting R&D in cellulosic biofuels, and Title XV, the tax and trade title, contains tax incentives and tariffs. The following programs provide support to cellulosic biofuels research, demonstration, and production.

Tax Credit for Cellulosic Biofuels (Section 15321)

Tax and trade provisions in the 2008 farm bill benefit cellulosic biofuels. One significant incentive is a production tax credit of $1.01 per gallon that applies to cellulosic biofuels production, more than twice the blenders' tax credit of 45 cents per gallon that applies to corn ethanol. In the case of cellulosic ethanol, the $1.01 credit amount is reduced by (1) the ethanol blender's credit and (2) the small ethanol producer credit, making the total value of tax incentives for cellulosic ethanol equal $1.01.[76]

Ethanol Tariff Extension (Section 15333)

In addition to tax credits, an ethanol tariff benefits the U.S. industry by reducing the competitiveness of imported ethanol sold in this country. Domestic ethanol benefits from a 54-cent-per-gallon duty (and a smaller *ad valorem* tariff) on imported ethanol (except for limited imports under the Caribbean Basin Initiative).[77] The original intent of the tariff was to prevent imported ethanol from benefitting from the U.S. tax credit.

Agricultural Bioenergy Feedstock and Energy Efficiency Research and Extension Initiative (Section 7207)

This new program awards competitive matching (up to 50%) grants for projects supporting on- farm biomass crop research and the dissemination of results to enhance the production of biomass energy crops and their integration with the production of bioenergy. It consists of elements of earlier initiatives that were moved to the research title (Title VII) in the 2008 farm bill. Discretionary funding of $50 million annually is authorized for FY2008 through FY20 12, subject to appropriations. Funding has not been appropriated for this program in FY2008, FY2009, or FY2010.

Biorefinery Assistance (Section 9003)

This initiative provides loan guarantees for the development, construction, and retrofitting of commercial-scale biorefineries and provides grants to help pay for the development and construction costs of demonstration-scale biorefineries. Loan guarantees are limited to $250 million per project, subject to the availability of funds. The program received mandatory funding of $320 million ($75 million for FY2009 and $245 million for FY2010) for commercial-scale biorefinery loan guarantees, and discretionary funding, subject to appropriations, of $150 million annually for FY2009 through FY2012 for both demonstration and commercial-scale biorefineries. This program was originally authorized by the 2002 farm bill, but received no funding from FY2002 through FY2007.

Of the five applications (worth over $1 billion) received in FY2009, USDA funded two. In January 2009, USDA funded the first cellulosic biofuels plant under the Biorefinery Program with a loan guarantee of $80 million. The recipient, Range Fuels Inc., in Soperton, GA, also received a $76 million grant from DOE for this plant. The biorefinery project will use wood chips to produce a projected output of 20 million gallons of ethanol annually, beginning in 2010. The second application under this program was approved on June 24, 2009, for a $25 million biodiesel plant. USDA anticipates receiving and reviewing applications in FY20 10, but not awarding funds until FY20 11, pending regulations.

Repowering Assistance (Section 9004)

The Repowering Assistance program encourages existing biorefineries to replace fossil fuels used to produce heat or power at their facilities by making payments for installing new systems that use renewable biomass, or to produce new energy from renewable biomass. The 2008 farm bill provided mandatory funding of $35 million in FY2009 to remain available until expended. The program is also authorized to receive $15 million in discretionary funding for each of FY2009 through FY2012, pending appropriations. On June 12, 2009, USDA issued a Notice of Funding Availability in the *Federal Register* to make $20 million in payments to biorefineries converting to renewable biomass.[78] The remaining $15 million of mandatory funding is expected to be available in FY20 10 pending regulations. There has been no appropriation of discretionary funding.

Bioenergy Program for Advanced Biofuels (Section 9005)

The Bioenergy Program is the lead program under Title IX providing support for the development of conversion technologies for cellulosic biofuels. It was originally established by Executive Order in 1999 and provided Commodity Credit Corporation (CCC) incentive payments to ethanol and biodiesel producers on the basis of yearly increases in production. Eligibility is now limited to producers of advanced biofuels. Eligible producers entering into a contract with USDA are paid based on quantity and duration of advanced biofuel production and on net renewable energy content of the advanced biofuel. Under the 2002 farm bill (P.L. 107-171), the Bioenergy Program received total funding of $426 million during FY2003 to FY2006 but no appropriations for FY2007 or FY2008. The 2008 farm bill provides a total of $300 million in mandatory funding for FY2009 to FY2012 ($55 million annually in FY2009 and FY2010, $85 million in FY2011, and $105 million in 2012), and also authorizes $25 million annually, subject to appropriations, from FY2009 to FY20 12.

The first Notice of Contract Proposals was published in the *Federal Register* on June 12, 2009.[79] Over 180 applications were received and 161 were deemed eligible. USDA anticipates making payments totaling $30 million to eligible biorefineries in early 2010. The remaining $25 million unexpended from FY2009 and the $55 million of mandatory funding for FY2010 will be available for payments in FY20 11 following publication of proposed regulations.[80]

Biomass Research and Development Initiative (Section 9008)

This program was originally authorized in the 2002 farm bill (P.L. 107-171) and is administered jointly by USDA and DOE. It supports research on and development and demonstration of biofuels and biobased products, and the methods, practices, and technologies for their production. The 2008 farm bill provides mandatory funding of $118 million for FY2009 to FY20 12 ($20 million for FY2009, $28 million for FY2010, $30 million for FY2011, and $40 million for FY2012). The farm bill also authorizes the appropriation of $35 million for each of FY2009 through FY2012. The program received $5 million in FY2002, $14 million for each of FY2003, FY2004, and FY2005, and $12 million for FY2006, and was not funded in FY2007 and FY2008. On January 30, 2009, USDA and DOE announced the FY2009 funding opportunity of $25 million ($20 million of mandatory funding from the 2008 farm bill and $5 million from DOE). Award recipients have been selected, but no announcement has been made.

Biomass Crop Assistance Program (Section 9011)

This is a new program intended to help farmers establish and produce crops for conversion to bioenergy and assist with the collection, harvest, storage, and transportation of eligible material for use in a biomass conversion facility that produces heat, power, biobased products, or advanced biofuels.[81] The program is implemented by the Farm Service Agency with support from other federal and local agencies and has mandatory CCC funding of such sums as necessary.

The program is implemented in two phases. On June 11, 2009, a Notice of Funds Availability was published in the *Federal Register* announcing the implementation of the 2009 collection, harvest, storage, and transportation (CHST) portion of the program.[82] As of October 23, 2009, 114 biomass conversion facilities have been qualified and listed as approved. Over 125 facility agreements and applications are pending at the FSA national office with another 72 in progress at FSA state offices. A total of $23.6 million was obligated in FY2009 and $514 million is requested for FY2010. The FSA has also issued a draft environmental impact statement[83] and is expected to implement the final phase of the program following the publication of a proposed rule in 2010.[84]

Forest Biomass for Energy (Section 9012)

Under this new program, USDA's Forest Service is authorized to conduct a comprehensive research and development program to use forest biomass for energy. Other federal agencies, state and local governments, Indian tribes, land-grant colleges and universities, and private entities are eligible to compete for program funds. No mandatory funding is available, but discretionary appropriations of $15 million annually for FY2009 to FY20 12 are authorized. This program has not yet been implemented or appropriated funding. Priority research projects include the following:

- the use of low-value forest biomass for energy from forest health and hazardous fuels reduction treatment;
- the integrated production of energy from forest biomass into biorefineries or other existing manufacturing;
- the development of new transportation fuels from forest biomass; and
- the improved growth and yield of trees for renewable energy production.

Energy Improvement and Extension Act of 2008 (P.L. 110-343)

The Emergency Economic Stabilization Act of 2008 (P.L. 110-343), which incorporates the Energy Improvement and Extension Act of 2008, contains tax and trade incentives for renewable energy production. Enacted on October 3, 2008, it expands federal benefits for renewable energy to fuels and processes that previously did not qualify and limits trade practices that benefit foreign producers but do not enhance U.S. energy independence.

Expansion of the Allowance for Cellulosic Ethanol Property (Division B, Section 201)

Previous federal tax law limited the eligibility for first-year bonus depreciation of cellulosic biofuels to facilities producing ethanol; those producing non-ethanol fuels from cellulosic feedstocks did not qualify for the allowance. P.L. 110-343 does not limit the allowance to any particular type of cellulosic fuel or production process. Taxpayers can immediately write off 50% of the cost of facilities that produce cellulosic biofuels if such facilities are placed in service before January 1, 2013.

Legislative Proposals

Congress has shown a strong interest in the development of biofuels in general and cellulosic biofuels in particular. Debate may continue on the appropriate level of incentives needed to jump start the industry. Perhaps the most critical emerging issue is the federal mandate for cellulosic biofuels under the RFS—and the industry's potential to meet that mandate. In the long term, Congress might also consider the ongoing level of government support that is appropriate for the cellulosic biofuels industry—considered by some to be essential, especially if the RFS is to be met. Others contend such support could distort market signals. The general level of support in the form of grants and loans has been determined in the 2008 farm bill but will be revisited as appropriations are considered. The cellulosic biofuels tax credit applies to fuel produced from 2009 through 2012 and extension of this credit could be the subject of debate during the 111th Congress. In addition, Congress has considered the definition of biofuels and biofuel feedstocks that qualify for federal incentives.

Legislative Changes in the RFS Volume

Citing the RFS and corn ethanol production as contributing to rising food prices and high input costs for livestock and poultry producers, some are calling for a reduction of the RFS. During the 110th Congress, S. 3031 was introduced in May 2008, to limit the corn-starch component of the RFS to 9 billion gallons compared with 15 billion under the current law. Opponents of the reduction claim it would set back efforts to reduce the nation's dependence on foreign oil and achieve environmental goals. Reducing the corn-starch component of the RFS would increase the importance of advanced fuels, primarily cellulosic biofuels, in meeting the mandate. Legislation to alter RFS volume requirements has not been introduced in the 111th Congress.

Expanding Biomass Eligible under the RFS

The definition of forest-based renewable biomass under the RFS is considered by some to be too restrictive because it limits eligible woody biomass to privately planted trees and tree residue from actively managed tree plantations, and slash[85] and pre-commercial thinnings from nonfederal forests.

The definition of renewable biomass specifically excludes biomass from federal forests. Some suggest that this exclusion eliminates much potential biomass and creates regional disparities. One-third of the 755 million acres of forest in the United States is owned by the federal government—and this acreage is concentrated in the western states. Likewise, the exclusion of private, naturally regenerated forests affects the northern and southeastern parts of the country where other feedstocks eligible under the RFS may not be as readily available. According to some, biomass extraction could become a powerful tool for improving federal land management.[86] Markets for small-diameter trees would enable a wider range of options for management of wildlife habitat, forest hydrology, hazardous fuels reduction, and pest infestations. These markets are not likely to appear if federal forests remain excluded from the RFS.

The House Committee on Agriculture Subcommittee on Conservation, Credit, Energy, and Research held hearings during July 2008 on producer eligibility under the RFS. The subcommittee heard from government officials, researchers, and producers who provided an update on the implementation of the RFS and shared concerns on barriers to eligibility for many agricultural producers. Subsequently, a Senate Energy and Natural Resources Committee field hearing on forest waste for biofuels was held in South Dakota on August 18, 2008.

During the 111th Congress, H.R. 2454 would allow for modification of the non-federal lands portion of the definition of "renewable biomass." Low-value materials are frequently removed during fire or disease reduction efforts or ecosystem health supporting activities. Waste materials such as wood waste and wood residues from private forests are also included.

Also during the 110th Congress, the House passed the Comprehensive American Energy Security and Consumer Protection Act (H.R. 6899). It contained a sense of Congress provision recommending a broad definition of renewable biomass to "encourage cellulosic biofuels ... produced from a highly diverse array of feedstocks, allowing every region of the country to be a potential producer of this fuel." No Senate action was taken. No similar action has been taken in the 111th Congress thus far.

TIME FRAME FOR CELLULOSIC BIOFUELS PRODUCTION

Estimates for commercial production of cellulosic biofuels vary widely. Some small-scale plants may come online this year.[87] However, the pace of plant construction falls short of DOE's stated goal to make cellulosic ethanol competitive as a mature technology by 2012.[88] Some analysts have predicted a growth trend for the cellulosic ethanol industry similar to that for corn-starch ethanol. However, there is a major difference between the two: the basic process for making cornstarch ethanol (fermentation) is thousands of years old, whereas that for cellulosic is very new.

The USDA Office of Energy and New Uses projects that cellulosic biofuels are not expected to be commercially viable on a large scale until at least 2015.[89] However, the cellulosic biofuel portion of the RFS mandate is set at 3 billion gallons by 2015, a substantial amount. In its January 2009 baseline, the Food and Agricultural Policy Research Institute (FAPRI) of the University of Missouri assumes cellulosic biofuel production will fall behind the RFS and, as a consequence, the mandate will be waived by EPA.[90] In an August 2009 baseline update, FAPRI projects cellulosic ethanol production in 2013 at 245 mgpy, about one-third of the 1 billion gallons in the RFS for that year.[91]

End Notes

[1] Cellulose is the structural component of the primary cell wall of green plants.

[2] For more information on ethanol, see CRS Report R40488, *Ethanol: Economic and Policy Issues*, by Randy Schnepf.

[3] For more information on the RFS, see CRS Report R40155, *Renewable Fuel Standard (RFS): Overview and Issues*, by Randy Schnepf and Brent D. Yacobucci.

[4] U.S. Environmental Protection Agency, "Regulation of Fuels and Fuel Additives: Changes to Renewable Fuel Standard Program; Final Rule," 75 *Federal Register*, March 26, 2010.

[5] Greenhouse gases include carbon dioxide, methane, and nitrous oxide (CO_2, CH_4, and N_2O respectively).

[6] For more information on the lifecycle analysis of greenhouse gas emissions under the RFS, see CRS Report R40460, *Calculation of Lifecycle Greenhouse Gas Emissions for the Renewable Fuel Standard (RFS)*, by Brent D. Yacobucci and Kelsi Bracmort.

[7] *Breaking the Link between Food and Biofuels*, Bruce A. Babcock, Briefing Paper 08-BP 53, July 2008.

[8] Preliminary draft of the feedstock and technology chapters from a Purdue University study on "Cellulosic Biofuels: Technology, Market, and Policy Assessment," September 7, 2009. This study is being conducted by Purdue University under contract with CRS and is supported by Joyce Foundation funding.

[9] *Growing and Harvesting Switchgrass for Ethanol Production in Tennessee*, Clark D. Garland, Tennessee Biofuels Initiative, University of Tennessee Institute of Agriculture, UT Extension SP-701a.

[10] Corn stover consists of the cob, stalk, leaf, and husk left in the field after harvest.

[11] Bioenergy Feedstock Information Network, Biomass Resources, http://bioenergy

[12] USDA Economic Research Service, *Ethanol Reshapes the Corn Market*, Amber Waves, April 2006, http://www.ers.usda.gov/AmberWaves/April06/Features/Ethanol.htm.

[13] Cellulosic Ethanol: A Greener Alternative, by Charles Stillman, June 2006, *http://www.clean houston.org/energy* features/ethanol2.htm.

[14] *Growing and Harvesting Switchgrass for Ethanol Production in Tennessee*, Clark D. Garland, Tennessee Biofuels Initiative, University of Tennessee Institute of Agriculture, UT Extension SP-701a.

[15] Ibid.

[16] "DuPont Danisco and University of Tennessee Partner to Build Innovative Cellulosic Ethanol Pilot Facility," press release, Nashville, TN, July 23, 2008.

[17] Science News, "Giant Grass Miscanthus Can Meet US Biofuels Goal Using Less Land Than Corn Or Switchgrass," *Science Daily*, August 4, 2008, *http://www.sciencedaily.com/releases/2008/07/080730155344.htm.*

[18] Ibid.

[19] *Perennial Bioenergy Feedstocks Report to Chairman Collin Peterson*—House Agriculture Committee, April 5, 2007, North Central Bio-economy Consortium.

[20] *Carbon-Negative Biofuels from Low-Input High-Diversity Grassland Biomass*, by David Tilman, Jason Hill, and Clarence Lehman, Science, December 8, 2006: vol. 314, p. 1598.

[21] *Biomass as a Feedstock for a Bioenergy and Bioproducts Industry: The Technical Feasibility of a Billion-Ton Annual Supply*, Environmental Sciences Division, Oak Ridge National Laboratory, 2005, http://feedstockreview.ornl.gov/pdf/billion_ton_vision p36.

[22] IAE Bioenergy, *Sustainable Production of Woody Biomass for Energy*, March 2002, http://www.energycommission.org/files/finalReport/IV.4.c%20-%20Cellulosic%20Ethanol%20Fact%20Sheet.pdf.

[23] "Fast-Growing Trees Could Take Root as Future Energy Source," press release of Purdue University Study funded by DOE, *http://www.purdue.edu/UNS/html4ever/2006/060823.Chapple.poplar.html.*

[24] *Producing Ethanol from Wood*, presentation by Alan Rudie, USDA Forest Service Forest Products Laboratory, Madison, WI, http://www.csrees.usda.gov/nea/plants

[25] DOE refinery feedstock estimates and CRS calculations into large round bales.

[26] *Analysis of the Efficiency of the U.S. Ethanol Industry 2007*, by May Wu, Center for Transportation Research Argonne National Laboratory Delivered to Renewable Fuels Association on March 27, 2008.

[27] *From Biomass to Biofuels*, National Renewable Energy Laboratory; NREL/BR-510-39436, August 2006, http://www.nrel.gov/biomass

[28] *The Economics of Biomass Collection, Transportation, and Supply to Indiana Cellulosic and Electric Utility Facilities*, Working Paper #08-03, by Sarah Brechbill and Wallace Tyner Purdue University, April 25, 2008.

[29] *Biomass Multi-Year Plan*, DOE Office of the Biomass Program, March 2006. http://www1.eere.energy pdfs/biomass_program_mypp.pdf.

[30] *Biomass as a Feedstock for a Bioenergy and Bioproducts Industry: The Technical Feasibility of a Billion-Ton Annual Supply*, Environmental Sciences Division, Oak Ridge National Laboratory, 2005, http://feedstockreview.ornl.gov/pdf/billion_ton_vision

[31] *Breaking the Link between Food and Biofuels*, Bruce A. Babcock, Briefing Paper 08-BP 53, July 2008, Center for Agricultural and Rural Development, Iowa State University, http://www.card.iastate.edu.

[32] *Ibid.*

[33] *Ibid.*

[34] For more information on biofuels and food supplies see CRS Report RL34474, *High Agricultural Commodity Prices: What Are the Issues?*, by Randy Schnepf.

[35] "DuPont Danisco and University of Tennessee Partner to Build Innovative Cellulosic Ethanol Pilot Facility," TN, July 23, 2008, http://www.utbioenergy.org/NR/rdonlyres/7D4527E3-00F7-4574-92B1-E6749851AA41/1262/072308FINALUTDDCEPressReleaseDDCELetteralllogos.pdf.

[36] *Biofuels Energy Program 2007*, DOE http://www1.eere.energy

[37] A mixture of hydrogen and carbon monoxide.

[38] J. D. Wright and C. G. d'Agincourt, "Evaluation of Sulfuric Acid Hydrolysis Processes for Alcohol Fuel Production," *Biotechnology and Bioengineering Symposium*, no. 14, (New York: John Wiley and Sons, 1984), pp 105- 123, *http://qibioenergy.wordpress* .com/¬2008/03/08/acid-hydrolysis/; P. C. Badger, "Ethanol From Cellulose: A General Review," in J. Janick and A. Whipkey, eds., *Trends in New Crops and New Uses* (ASHS Press, 2002).

[39] DOE, EERE, Biomass Program, "Cellulase Enzyme Research," at http://www1.eere.energy biomass_feedstocks.html.

[40] *Development of New Sugar Hydrolysis Enzymes: DOE, Novozymes Biotech, Inc*. http://search.nrel.gov/cs.html?url= http%3A//www1 .eere. energy cellulase+enzymes&col=eren&n=2&la=en.

[41] *Biomass Multi-Year Plan*, DOE Office of the Biomass Program, March 2006. http://www1.eere.energy pdfs/biomass_program_mypp.pdf.

[42] *Biomass Multi-Year Plan*, DOE Office of the Biomass Program, March 2006. http://www1.eere.energy pdfs/biomass_program_mypp.pdf.

[43] J. D. Wright, "Evaluation of Sulfuric Acid Hydrolysis Processes for Alcohol Fuel Production," in *Biotechnology and Bioengineering Symposium*, no. 14, (New York: John Wiley and Sons, 1984), pp 103-123.

[44] P. C. Badger, "Ethanol From Cellulose: A General Review," in J. Janick and A. Whipkey, eds., *Trends in New Crops and New Uses* (ASHS Press, 2002)..

[45] *Biomass Multi-Year Plan*, DOE Office of the Biomass Program, March 2006. http://www1.eere.energy pdfs/biomass_program_mypp.pdf.

[46] For more information on the blend wall, see CRS Report R40445, *Intermediate-Level Blends of Ethanol in Gasoline, and the Ethanol "Blend Wall"*, by Brent D. Yacobucci.

[47] E-10 refers to a fuel blend of 10% ethanol and 90% gasoline. Likewise, E-15 is a blend of 15% ethanol, 85% gasoline; E-20 is 20% ethanol, 80% gasoline; and E-85 is 85% ethanol, 15% gasoline.

[48] Potential drop-in fuels include synthetic gasoline or diesel fuel produced from biomass, as well as butanol or other chemicals that may not have some of the blending limitations faced by ethanol.

[49] Environmental Protection Agency, "EPA Grants E15 Waiver for Newer Vehicles/A New Label for E15 Is Being Proposed to Help Ensure Consumers Use the Correct Fuel," press release, October 13, 2010, http://yosemite.epa.gov/ opa/admpress.nsf/0/BF822DDBEC29C0D C852577BB005BAC0F.

[50] *Research Advances in Cellulosic Ethanol*, DOE-NREL, at http://www.nrel.gov/biomass/ March 2007.

[51] *The Energy Balance of Corn Ethanol: An Update*, Shapouri, Hosein; James A. Duffield, and Michael Wang. USDA, Office of the Chief Economist, Office of Energy Policy and New Uses, Agricultural Economic Report (AER) No. 813, July 2002; available at http://www.usda.gov/oce/reports/energy

[52] *Cellulosic Ethanol Fact Sheet*, by Lee R. Lynd, presented at the National Commission on Energy Policy Forum: The Future of Biomass and Transportation Fuels, June 13, 2003.

[53] Worldwatch Institute, *Biofuels for Transportation, Global Potential and Implications for Sustainable Agriculture and Energy in the 21st Century*. Table 10-1, p. 127, June 2006.

[54] *Ethanol From Biomass: Can It Substitute for Gasoline?* Michael B. McElroy, book chapter draft.

[55] *Net Energy of Cellulosic Ethanol from Switchgrass*, M.R. Schimer, K.P. Vogel, R.B. Mitchell, and R.B. Perrin, PNAS January 15, 2008, vol. 105, no. 2, available at http://www.pnas.org/cgi/doi/10.1073/pnas.0704767105.

[56] Timothy A. Volk, Theo Verwijst, and Pradeep J. Tharakan et al., "Growing Fuel: A Sustainability Assessment of Willow Biomass Crops," *Frontiers in Ecology & Environment Journal*, vol. 2, no. 8 (2004), pp. 411-418.

[57] A herbaceous plant is a plant that has leaves and stems that die down at the end of the growing season to the soil level.

[58] Wang, et al., table 7.

[59] *Greenhouse Gas Impacts of Expanded Renewable and Alternative Fuels Use*, EPA Office of Transportation and Air Quality, EPA-420-F-07-035, April 2007.

[60] *BP and Verenium Partner To Commercialize Cellulosic Ethanol*, RenewableEnergy-World.com, August 11, 2008, *http://www.renewableenergyworld.com/rea/news/story?id=53280.*

[61] *Shell Boosts Second Generation Biofuels*, by Ed Crooks, Financial Times, July 16, 2008, http://www.ft.com/cms/s/ 7f9f1ee8-52d0-1 1dd-9ba7-000077b07658.html.

[62] *U.P. Biofuel Plant Lands $50m in State, Fed Aid*, by Gary Heinlein and David Shepardson, The Detroit News, October 8, 2008, *http://www.detnews.com/apps/pbcs.dll/article?AID=/20081008/BIZ/810080393.*

[63] *Monsanto Company and Mendel Biotechnology, Inc. Announce Cellulosic Biofuels Collaboration*, BioSpace.com, April 28, 2008, *http://www.biospace.com/news_story.aspx?NewsEntityId=94113.*

[64] "2007 State of the Union Address" 20 in 10: *Strengthening America's Energy Security*, http://www.whitehouse.gov/ stateoftheunion/2007/initiatives/energy.html.

[65] For more information on federal biofuels incentives see CRS Report R401 10, *Biofuels Incentives: A Summary of Federal Programs*, by Brent D. Yacobucci.

[66] The National Biofuels Action Plan is available at http://www.eere.energy

[67] FY2009 DOE Budget Request to Congress, available at *http://www.cfo.doe.gov/budget/09budget/* start.htm#Detailed%20Budget%20Justifications.

[68] Presentation at Platts Cellulosic Ethanol Conference, by Valri Lightner, DOE Biomass Program, October 2008, http://www.autobloggreen.com/photos/platts-conference

[69] CRS Report R40669, *Energy and Water Development: FY2010 Appropriations*, coordinated by Carl E. Behrens.

[70] U.S. Congress, House, *Energy and Water Development and Related Agencies Appropriations Act, 2010*, conference report to accompany H.R. 3183, 111th Cong., 1st sess., September 30, 2009, H.Rept. 111-278, pp. 282, 283, 290.

[71] The values in this paragraph are from personal correspondence with USDA's Office of Budget and Policy Analysis.

[72] "Iogen Suspends U.S. Cellulosic Ethanol Plant Plans," Earth2tech, http://earth2tech.com-/2008/06/04/iogensuspends-us-cellulosic-ethanol

[73] "Alico to Discontinue Ethanol Efforts," June 2, 2008, press release, http://www.aliconinc.-com/june208.asp.

[74] CRS Report RL34 130, *Renewable Energy Programs in the 2008 Farm Bill*, by Megan Stubbs

[75] Title IX cites are amendments to the 2002 farm bill (P.L. 107-17 1).

[76] For more information, see CRS Report R401 10, *Biofuels Incentives: A Summary of Federal Programs*, by Brent D. Yacobucci.

[77] For more information about ethanol imports under the CBI, see CRS Report RS2 1930, *Ethanol Imports and the Caribbean Basin Initiative (CBI)*, by Brent D. Yacobucci.

[78] USDA, Rural Business-Cooperative Service, "Notice of Funding Availability for Repowering Assistance Payments to Eligible Biorefineries," 74 *Federal Register* 112, June 12, 2009.

[79] Rural Business-Cooperative Service, USDA, "Notice of Contract Proposal for Payments to Eligible Advanced Biofuel Producers," 74 *Federal Register* 112, June 12, 2009.

[80] U.S. Congress, House Committee on Agriculture, Subcommittee on Conservation, Credit, Energy, and Research, *Statement of Dallas Tonsager, Under Secretary for Rural Development, USDA*, hearing to review the next generation biofuels, 111th Cong., 1st sess., October 29, 2009, p. 6.

[81] For more information, see CRS Report R41296, *Biomass Crop Assistance Program (BCAP): Status and Issues*, by Megan Stubbs.

[82] Commodity Credit Corporation, USDA, "Notice of Funds Availability for the Collection, Harvest, Storage, and Transportation of Eligible Material," 74 *Federal Register* 11, June 11, 2009.

[83] Commodity Credit Corporation, USDA, "Notice of Availability of the Draft Programmatic Environmental Impact Statement for the Biomass Crop Assistance Program," 74 *Federal Register* 152, August 10, 2009.

[84] Office of Budget and Program Analysis, *Highlights: Farm Bill Working Group*, USDA, Title IX - Energy, October 26, 2009.

[85] The accumulation of limbs, tops, and miscellaneous residue left by forest management activities, such as thinning, pruning, and timber harvesting.

[86] *Federal Forests and the Renewable Fuel Standard Factsheet*, Environmental and Energy Study Institute, July 17, 2008.

[87] For more information, see CRS Report R41 106, *Meeting the Renewable Fuel Standard (RFS) Mandate for Cellulosic Biofuels: Questions and Answers*, by Kelsi Bracmort.

[88] *Biomass Multi-year Program Plan*, DOE, March 2008, http://www1.eere.energy publications.html#vision.

[89] *U.S. Biobased Products: Market Potential and Projections Through 2025*, Office of the Chief Economist, Office of Energy Policy and New Uses, USDA, the Center for Industrial Research and Service of Iowa State University, Informa Economics, Michigan Biotechnology Institute, and The Windmill Group. OCE-2008-1, http://www.oce.usda.gov.

[90] FAPRI, Iowa State University, and University of Missouri-Columbia, *FAPRI 2009: U.S. and World Agricultural Outlook*, Report 09-FSR 1, January 2009.

[91] *Baseline Update for Agricultural Markets*, FAPRI, FAPRI MU-Report #06-09, August 2009. http://www.fapri.missouri.edu.

In: Cellulosic Biofuels: Feedstock, Technology… ISBN: 978-1-61209-509-7
Editors: Robert M. Burkin et al.

Chapter 3

MEETING THE RENEWABLE FUEL STANDARD (RFS) MANDATE FOR CELLULOSIC BIOFUELS: QUESTIONS AND ANSWERS

Kelsi Bracmort

SUMMARY

The Renewable Fuel Standard (RFS) was expanded under the Energy Independence and Security Act of 2007 (EISA; P.L. 110-140) in an effort to reduce dependence on foreign oil, promote biofuel use, and stabilize transportation fuel prices, among other goals. Over a 15-year period, the RFS seeks to establish a market for biofuels in the transportation sector by requiring that increasing amounts of biofuels-36 billion gallons by 2022—be blended into transportation fuel. The mandate is to be accomplished with an assortment of advanced biofuels, including cellulosic biofuels—fuels produced from cellulosic materials including grasses, trees, and agricultural and municipal wastes. The cellulosic biofuel allotment in the mandate, as established by Congress in EISA, was 100 million gallons due in 2010, increasing to 16 billion gallons by 2022. However, on March 26, 2010, the U.S. Environmental Protection Agency (EPA) issued a fmal rule for implementation of the RFS that sets a new, lower cellulosic biofuel mandate of 65 million gallons for 2010.

Recent analysis has suggested that the United States might not have sufficient cellulosic biofuel production capacity to meet the 2010 RFS mandate of 100 million gallons instituted by Congress in EISA. The cellulosic biofuel community may fare better at achieving the new mandate set by EPA if certain obstacles are overcome. No commercial-scale cellulosic biofuel plants are currently operating. Roadblocks include unknown levels of feedstock supply, expensive conversion technology that has not yet been applied commercially, and insufficient financial support from private investors and the federal government.

Some financial support from the Departments of Energy and Agriculture is available to expedite cellulosic biofuel production. For example, the Biomass Crop Assistance Program (BCAP), created under the Food, Conservation, and Energy Act of 2008 (2008 farm bill; P.L. 110-246), is to support establishment and production of crops for conversion to bioenergy. Financial support available thus far via BCAP is for collection, harvest, storage, and transportation of eligible material. BCAP support for the establishment and production of eligible crops for the conversion to bioenergy is anticipated to begin in 2010. Also, the Department of Energy's Loan Guarantee Program, created under the Energy Policy Act of 2005 (EPAct05, P.L. 109-58), distributes loan guarantees to eligible commercial-scale renewable energy systems, including cellulosic biofuel plants, although criticisms have been raised that the program has been slow to get started.

Many questions regarding cellulosic biofuels and the RFS may arise as Congress engages in energy legislation debates. Can and will the 2010 and future RFS mandates for cellulosic biofuels be met? What impact will significantly lowering the 2010 cellulosic ethanol mandate have on investment in celluosic ethanol production? What are the next steps Congress could take to expedite cellulosic biofuel production? Proposed legislation (H.R. 2454, S. 1462, H.R. 2283, and S. 943), if enacted, may influence cellulosic biofuel production by providing additional financial, infrastructure, and environmental support. This chapter, in a question and answer format, discusses some of the concerns facing the cellulosic biofuel community, including feedstock supply estimates, an expected time frame for the first commercial cellulosic biofuel projects, and potential legislative options to address cellulosic biofuel production uncertainty for the RFS.

INTRODUCTION

The Renewable Fuel Standard (RFS) was enacted to ensure that domestic transportation fuel contains a specified volume of biofuels, including advanced biofuel, cellulosic biofuel, and biomass-based diesel.[1] The RFS requires that increasing amounts of biofuels be blended into U.S. transportation fuel over a 15-year period, with the goal of using 36 billion gallons of biofuels annually by 2022. The mandate is to be accomplished in large part with cellulosic biofuels. The cellulosic biofuel allotment in the mandate established by Congress began with 100 million gallons due in 2010, increasing to 16 billion gallons by 2022. However, on March 26, 2010, the U.S. Environmental Protection Agency (EPA) issued a final rule for the RFS lowering the 2010 cellulosic biofuel mandate to 6.5 million gallons.[2]

A concern for Congress is whether enough cellulosic biofuel can be produced to satisfy the RFS mandate in future years. Cellulosic feedstock supply, financial assistance, and technology advancement are considered among the most pressing issues that could thwart cellulosic biofuel production. More than two years after EISA was enacted, progress towards meeting the cellulosic biofuels mandate has been delayed on multiple fronts (e.g., fmancial, administrative, and technical), despite government financial incentives, estimates of feedstocks available, and technological developments. If the revised 2010 RFS cellulosic biofuel mandate is not met, Congress may reconsider the configuration of the RFS, determine whether additional resources are necessary for cellulosic biofuel production, and assess the success rate of this effort compared to other renewable energy efforts.

WHAT ARE CELLULOSIC BIOFUELS?

Cellulosic biofuels are liquid, solid, or gaseous fuels made from cellulose material. Cellulose—a complex carbohydrate—is the organic matter found in plant walls that, along with hemicellulose and lignin, helps to give a plant its rigid structure. Cellulose feedstock includes agricultural residues (e.g., corn stover), forestry residues (e.g., wood chips), energy crops (e.g., switchgrass), tree crops (e.g., hybrid poplar), and urban sources of waste (e.g., municipal solid waste).

The most widely discussed cellulosic biofuel is cellulosic ethanol for transportation.[3] Cellulosic ethanol differs from the corn ethanol currently blended into transportation fuel; it is made from feedstock with no food value, potentially results in fewer greenhouse gas emissions, and has a higher energy balance.[4] Converting cellulosic feedstock to ethanol, however, is more expensive and difficult than converting corn to ethanol. The conversion of cellulose to ethanol generally happens in three phases—pretreatment, hydrolysis, and fermentation to ethanol. Pretreatment weakens the plant wall structure so that the cellulose is easier to obtain during hydrolysis.

Hydrolysis—acid or enzymatic—separates the cellulose into sugars. Fermentation converts the sugars into ethanol. Cellulose can also be converted to liquid fuels through processes other than fermentation (e.g., thermochemical processes).[5]

Analysis suggests that increased use of cellulosic biofuels for transportation could potentially help to reduce U.S. dependence on foreign oil, stabilize energy prices, strengthen rural infrastructure, and improve the environment. In addition, cellulosic feedstocks may fare better in the food-energy debate, since crop residue, and not the crop itself, is used for cellulosic biofuel production. Some contend, however, that cellulosic biofuels require a substantial feedstock supply that has yet to be verified, may cause environmental degradation (e.g., by removing residues that furnish nutrients and stability to the soil),[6] and may hinder efforts to promote energy efficiency.

What Is the Relationship between Cellulosic Biofuels and the Renewable Fuel Standard?

The RFS established in Section 202 of EISA called for 100 million gallons of cellulosic biofuels to be blended in the national transportation fuel supply in 2010, and the mandate increases to 16 billion gallons by 2022.[7] Data and analysis presented during the RFS debate and ultimate passage of EISA in 2007 supported these levels of cellulosic biofuel production capacity. Some reasoned that plentiful feedstock was available[8] and that the conversion technology was on the brink of being certified as commercially viable. Moreover, some presumed that the federal government would provide substantial financial support and enhance the infrastructure needed to spur a commercial cellulosic biofuels market.[9] Others were leery about the time frame provided to meet the RFS cellulosic biofuel mandate given capacity restrictions, weather impacts, and uncertainty about technology advancements.[10]

EPA has the authority to waive completely or in part the 100 million gallons per year (mgy) 2010 mandate established in EISA, given certain circumstances.[11] For instance, the Administrator may waive the cellulosic biofuel requirement if the Administrator determines, after public notice and opportunity for comment, that there is an inadequate domestic supply.[12] On March 26, 2010, EPA lowered the 2010 cellulosic biofuel mandate set forth in EISA by roughly 93% with the issuance of a waiver that expires after one year:

> EISA requires the Administrator to evaluate and make an appropriate market determination for setting the cellulosic standard each year. Based on an updated market analysis considering detailed information from pilot and demonstration scale plants, an Energy Information Administration analysis, and other publically and privately available market information, we are setting the 2010 cellulosic biofuel standard at 6.5 million ethanol-equivalent gallons. While this volume is significantly less than that set forth in EISA for 2010, a number of companies and projects appear to be poised to expand production over the next several years.°

The 2011 renewable fuel standard for cellulosic biofuels was 250 millions gallons. However, EPA announced on July 9, 2010, that it would lower this mandate to a range of 5-17 1 million gallons.[14] For the final rule, EPA intends to pick a single value from within this range.

What Challenges Are Associated with Cellulosic Biofuels Production?

Over the two years since the RFS was expanded,[15] obstacles have stalled U.S. cellulosic biofuel production capacity. Impediments to meeting the cellulosic biofuel mandate include technology setbacks, escalating prices for certain feedstocks, lack of feedstock availability, and delayed financial support. Limited access to capital has been indicated as one of the primary reasons that timely completion of many cellulosic biofuel plants has stalled. Commercial cellulosic biofuel facilities are estimated to cost hundreds of millions of dollars (see **Table 1**), roughly three times as much as a corn ethanol plant. (In comparison, a 40 mgy corn ethanol plant costs approximately $80 million to construct.[16]) Some lenders find it extremely risky, perhaps even cost-prohibitive, to provide financial backing to cellulosic biofuel plants, mainly because the conversion technology has not been applied or proven on a large scale.

The government provides financing through two programs to encourage investment in cellulosic biofuel production technologies. In addition, the 2008 farm bill offers a production credit of $1.01 per gallon for biofuels produced from qualifying cellulosic feedstocks.[17] To increase investment, the government established the Department of Energy (DOE) Loan Guarantee Program (LGP).[18] Loans may not exceed 80% of total project costs. Over 90% of the projects that have received funding to date are pilot- or demonstration-scale projects that are seen as likely to become a commercial technology.[19] Biomass renewable energy technologies accounted for approximately 24% of applications for the LGP in 2008. DOE received $18.5 billion toward the LGP for energy- efficiency and renewable energy projects—including cellulosic biofuel projects—in the 2009 Omnibus Appropriations Act (P.L. 111-8).[20] An additional $60 billion in loan guarantees was provided under the American Recovery and Reinvestment Act of 2009 (ARRA, P.L. 111-5), partly for renewable energy projects.[21] Some are concerned that the LGP is not being carried out at a pace responsive to market momentum for cellulosic biofuels.[22] Recently, DOE distributed roughly $564 million to 19 advanced biorefinery projects, including cellulosic ethanol plants.[23]

The second government program is the U.S. Department of Agriculture's (USDA's) Biomass Crop Assistance Program (BCAP).[24] BCAP is to assist "agricultural and forest land owners and operators with matching payments for the amount paid for the collection, harvest, storage and transportation (CHST) of eligible material by a qualified Biomass Conversion Facility (BCF)" and to support "establishing and producing eligible crops for the conversion to bioenergy through project areas and on contract acreage up to 5 years for annual and non-woody perennial crops or up to 15 years for woody perennial crops." The CHST portion of the program began in June 2009. USDA's Farm Services Agency (FSA) anticipates implementing the bioenergy crops portion of the program in FY2010. Some question the intent of BCAP and whether it is effective at assisting the groups Congress originally intended.[25]

In addition to financing issues, other challenges to cellulosic biofuel production include the definitions of biomass in various laws. The renewable biomass definition for the RFS under EISA does not allow biomass removal from federal lands, and the law excludes crops from forested lands.[26] Some argue that opening up federal lands for biomass removal could provide an inexpensive supply of cellulosic feedstock that would be immediately available to biorefineries for cellulosic biofuel production. Others contend that biomass removal from federal lands is a short-term response to the cellulosic feedstock source problem and might not be carried out in a sustainable manner, leading to deterioration of the nation's parks and recreation areas. The definition of biomass under EISA excludes most municipal solid waste (MSW), which some view as a potential source for conversion to biofuels.

Also challenging for cellulosic biofuel production are the time periods of feedstock contracts. Agricultural and forestry producers may not agree to a contract that requires a lengthy time commitment. For example, it generally takes three years for switchgrass crops to reach maturity.[27] A producer may have to commit his land to one particular cellulosic feedstock crop for a number of years, thus limiting the producer's choice to grow certain crops on an annual basis depending upon market demand.

Table 1. Selected Examples of Cellulosic Ethanol Plant Cost Estimates

Company	Production Capacity (mgy)	Feedstock Required	Capital (millions)
BlueFire (California Plant)a	3.9	190 wet tons/day of post-sorted municipal solid waste	$120
BlueFire (Mississippi Plant)[a]	19	550 tons/day of wood waste (mostly forest residues)	$250
Enerkem (Mississippi Plant)[b]	10	300 dry tons/day of post-sorted municipal solid waste	$118
POET (Iowa Plant)[c]	25	770 dry tons/day (mostly corncobs with some corn residue)	$200
ZeaChem Inc. (Oregon Plant)[d]	0.25	10 dry tons/day of hybrid polar trees	$73

Source: Compiled by CRS.

a. Conversation with Arnold Klann from BlueFire Ethanol Inc., February 2, 2010.

b. E-mail from Marie-Helene Labrie of Enerkem, February 3, 2010.

c. Conversation with Jim Sturdevant from POET, February 2, 2010.

d. Conversation with Carrie Atiyeh from ZeaChem Inc., February 2, 2010.

WILL THE REVISED 2010 RFS MANDATE FOR CELLULOSIC BIOFUELS BE MET?

Even before enactment of EISA, reported production data indicated that overcoming any or all of the hurdles to increase cellulosic biofuel production to meet the original 2010 RFS mandate of 100 million gallons set by Congress was unlikely. The new 2010 cellulosic biofuel mandate of 6.5 million gallons announced by EPA may be attainable with a relatively small number of pilot- or demonstration-scale projects. EPA acknowledges that the pilot or demonstration facilities tasked with producing the majority of the cellulosic biofuel mandate seldom operate continuously or at full "nameplate" capacity. Thus, some argue, it is difficult to state with certainty how much cellulosic biofuel will be produced and over what time frame. EPA anticipates that more than one- third of the mandate will be met with cellulosic ethanol production from the Range Fuels plant located in Georgia.[28]

According to data provided by the Renewable Fuels Association, two of the 20 U.S. cellulosic ethanol projects currently under construction or development that could produce more than 1 mgy are expected to come online in 2010.[29] Combined, these two facilities are expected to produce roughly 18 mgy of cellulosic ethanol in their first stage of operation. Cellulosic ethanol production capacity of 100 million gallons may not be reached until 2012 or later.[30]

WHAT IMPACT WILL SIGNIFICANTLY LOWERING THE 2010 RFS MANDATE HAVE ON INVESTMENT IN CELLULOSIC BIOFUEL PRODUCTION?

As noted above, EPA has the authority to waive the cellulosic biofuel mandate on a yearly basis. Indeed, EPA issued a waiver to substantially lower the 2010 cellulosic biofuel mandate and proposes to do the same for 2011. EPA's waiver authority creates uncertainty for investors in cellulosic biofuel ventures. Investors may fear that the cellulosic biofuel mandate will never be required by EPA. Some lenders may deny financing due to lack of confidence in federal support for cellulosic biofuels.[31] If the cellulosic biofuel community is unable to produce 10% of the cellulosic biofuel mandate, as established in EISA, for its first year of inclusion in the RFS, some may wonder about the viability of this advanced biofuel type over the long term.[32] Others contend the cellulosic biofuel community hit a stumbling block mainly due to a bad economy, and that production will drastically increase over the coming years.[33]

HOW MUCH CELLULOSIC FEEDSTOCK EXISTS FOR CONVERSION TO BIOFUELS?

A significant criterion in evaluating whether a commercial cellulosic biofuel production plant will be a favorable investment is whether a steady feedstock supply exists at a location suitable to the biorefinery.[34] However, determining actual availability of feedstock is difficult. Quantifying feedstock available for conversion to biofuels requires information about feedstock sources, production rates, accessibility, and location restrictions (e.g., public versus private lands if the feedstock is to be used for certain energy purposes). Investors must make feedstock predictions based on data from weather patterns and land use change, as well as handling, storage, and transportation costs, among other things. This is a particularly important problem where a growth season of four to five months must provide biomass feedstock for 12 months of plant operation. To make the supply available throughout the year, special equipment may be required for feedstock harvest, handling, collection, storage, or transport, as cellulosic feedstock is often too bulky for average farming equipment to handle.

The amount of cellulosic feedstock necessary for conversion to a biofuel depends on the feedstock type, the conversion process, and the desired biofuel (see **Table 2**). Biofuel conversion yield is measured in gallons per ton. Feedstock, or crop, yield is measured in tons per acre. Total yield, measured in gallons per acre, depends on both the feedstock yield and the conversion yield. Some argue that current estimated cellulosic feedstock yields will need to increase markedly over the next decade to meet the RFS mandate of 16 billion gallons of cellulosic biofuel production per year by 2022. Others contend that a significant growth of cellulosic feedstock is not essential, as advances in conversion technologies will afford the opportunity to produce more cellulosic biofuel with less feedstock.

If cellulosic feedstock yields do increase, the traditional geographic areas for feedstock cultivation may be subject to additional energy, environmental, and agricultural policy scrutiny (see **Figure 1**).

Table 2. Theoretical Ethanol Production Yields for Selected Feedstocks

Feedstock	Ethanol (gallons per dry ton of feedstock)
Corn Grain	124.4
Corn Stover	113.0
Rice Straw	109.9
Cotton Gin Trash	56.8
Forest Thinnings	81.5
Hardwood Sawdust	100.8
Bagasse	111.5
Mixed Paper	116.2

Source: U.S. Dept. of Energy, Biomass Program, http://wwwl.eere.energy ethanol_yield_calculator.html.

Notes: Actual yield commonly ranges from 60% to 90% of theoretical yields.

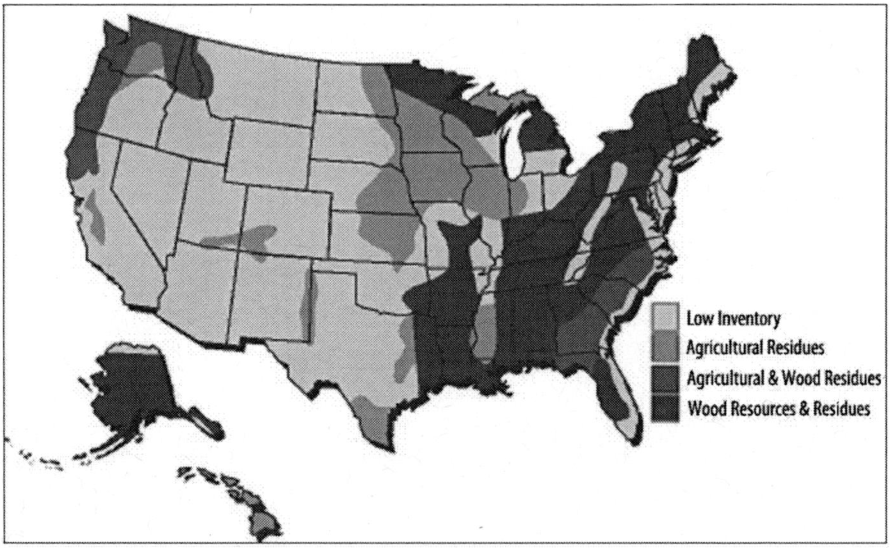

Source: U.S. Department of Energy, http://www.energysavers. gov/images

Figure 1. Geographical Areas with Biomass Resources

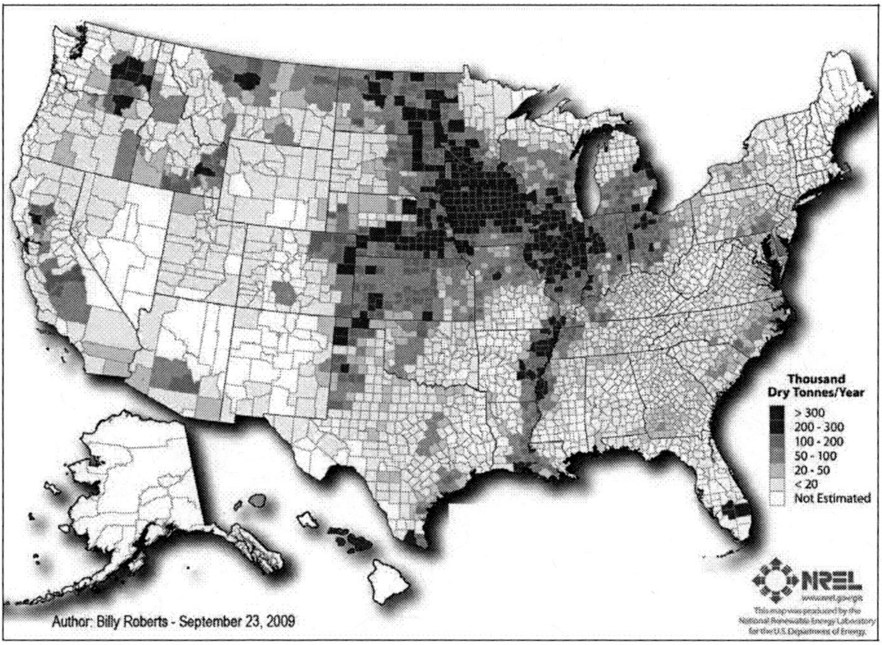

Source: U.S. Department of Energy, National Renewable Energy Laboratory, http://www.nrel.govigis/images map_biomass_crop_residues.jpg. Feedstock estimate based on five-year average for 2003 to 2007 from the U.S. Department of Agriculture, National Agricultural Statistics Service.

Note: The following crops were included in this analysis: corn, wheat, soybeans, cotton, sorghum, barley, oats, rice, rye, canola, dry edible beans, dry edible peas, peanuts, potatoes, safflower, sunflower, sugarcane, and flaxseed. The quantities of crop residues that can be available in each county are estimated using total grain production, crop to residue ratio, and moisture content, taking into consideration the amount of residue left on the field for soil protection, grazing, and other agricultural activities. USDA, National Agricultural Statistics Service, five-year average, 2003-2007 data.

Figure 2. Cellulosic Feedstock Available from Crop Residues in the United States

Few studies have estimated the current or long-term cellulosic feedstock supply available for conversion to biofuels on a national basis. Estimates of the amount of cellulosic feedstock available in recent years (2003-2007) range from 5,000 to 300,000 tons per year (**Figure 2** and **Figure 3**). Assuming a yield of 100 gallons per ton, this would equate to between 0.5 and 30 mgy of potential fuel production from that feedstock. This is well below the 100 million gallons needed to meet the 2010 mandate originally set by Congress, but is within range of the 6.5 million gallons production volume set by EPA for 2010.

Long-term cellulosic feedstock supply projections—for varying time periods beginning in 2020 and continuing to 2050—range from roughly 500 million to more than 1 billion tons per year capable of producing biofuels, well beyond the 16 billion gallons needed to meet the 2022 mandate **(Table 3).**[35]

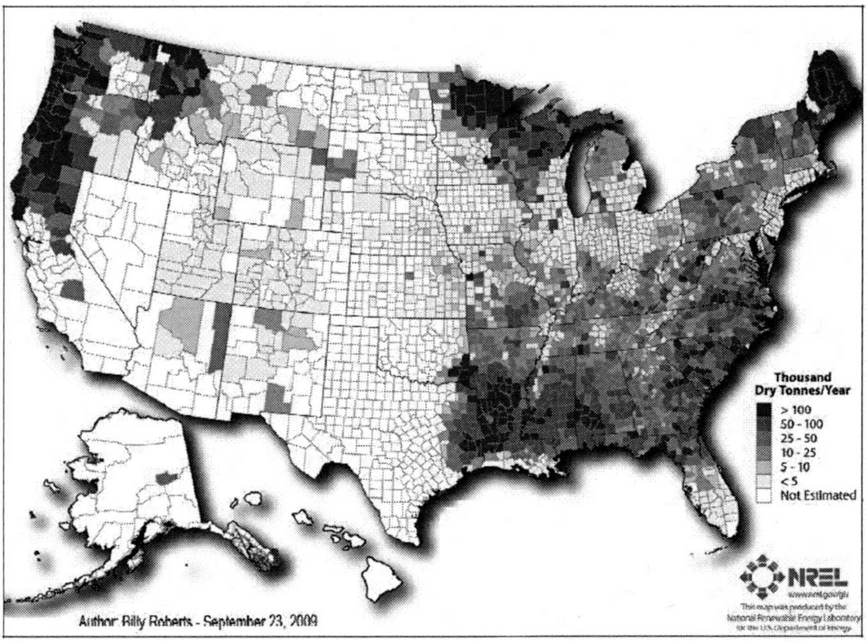

Source: U.S. Department of Energy, National Renewable Energy Laboratory, http://www.nrel.govigis/images map_biomass_forest_residues.jpg. Feedstock estimate based on U.S. Forest Service Timber Product Output database for 2007.

Note: Forest residues include logging residues and other removable material left after carrying out silvicultural operations and site conversions. Logging residue comprises unused portions of trees cut or killed by logging and left in the woods. Other removable materials are the unutilized volumes of trees cut or killed during logging operations. USDA, Forest Service's Timber Product Output database, 2007.

Figure 3. Cellulosic FeedstockAvailable from Forest Residues in the United States

Table 3. Cellulosic Feedstock Supply Estimates of Availability and Need.

Source	Feedstock Quantity (Tons)	Comments
Estimates of Availability		
USDA / DOE Billion-Ton Study (2005)[a]	1.366 billion available on a yearly basis roughly around the mid-21st century • 368 million from forest lands (27% of estimated total) • 621 million from agricultural lands[b] (45% of estimated total) • 377 million for dedicated energy crops[c] (28% of estimated total)	The agricultural land component of the estimated total includes grains (corn and soybean) used for ethanol and biodiesel which are not cellulosic feedstock. The study included biomass available for bioenergy in general (e.g., wood for electricty). The Billion-Ton Study may have overestimated the amount of feedstock that can be economically harvested because it did not calculate costs associated with harvesting potential feedstocks using existing technology. The study also included woody biomass from federal forest lands, but EISA subsequently excluded such biomass from qualifying under the RFS. An updated study is expected to be published later this year.[d]
National Academy of Sciences (2009)[e]	548 million available in 2020 • 124 million from forest lands (23% of estimated total) • 160 million from agricultural lands[f] (29% of estimated total) • 164 million from dedicated energy crops (30% of estimated total)	The study estimates that the cellulosic feedstock can be available sustainably for conversion to liquid fuel in 2020.
Estimates of Need		
Sandia National Laboratories / General Motors' R&D Center (2009)[g]	775 million • 126 million tons from forest lands (16% of estimated total) • 182 million from agricultural lands (23% of estimated total) • 467 million from dedicated energy crops (60% of estimated total)	The study estimates that 775 million tons of biomass are required to produce 70 billion gallons/year (BGY) of cellulosic ethanol by 2030

Table 3. (Continued)

Source	Feedstock Quantity (Tons)	Comments
Biomass Research and Development Initiative (2008)[h]	240 million • 153 million from agricultural lands[i] (64% of estimated total) • 85 million from dedicated energy crops (35% of estimated total)	The study estimates, for one scenario out of the three provided, that 240 million tons of biomass are required to produce 36 BGY of cellulosic ethanol by 2022. Assumptions for the scenario include meeting the 36 BGY of cellulosic ethanol with corn-based ethanol of 15 BGY, soybean diesel of 1 BGY, 20 BGY from cropland biomass, 0 BGY from forestland biomass, and 0 BGY from imports.

Source: Compiled by CRS.

a. DOE, USDA, *Biomass as Feedstock for a Bioenergy and Bioproducts Industry. The Technical Feasibility of a Billion- Ton Annual Supply*, April 2005, http://www I .eere.energy.gov/biomass/pdfs/ final_billionton_vision_report2.pdf.

b. Agricultural lands include crop residues, grains, and animal manure.

c. Perennial crops (e.g., grasses) are generally considered dedicated energy crops.

d. U.S. Government Accountability Office, *Biofuels: Potential Effects and Challenges of Required Increases in Production and Use*, GAO-09-446, August 2009, http://www.gao.gov/new.items/d09446.pdf.

e. National Academy of Sciences, National Academy of Engineering, National Research Council, *Liquid Transportation Fuels from Coal and Biomass: Technological Status, Costs, and Environmental Impacts*, Washington, DC, 2009, *http://www.nap.edu/ catalog.php?record_id=* 12620.

f. Agricultural lands include corn stover, wheat and grass straw, hay, and animal manure.

g. Sandia National Laboratories, *90-Billion Gallon Biofuel Deployment Study*, February 2009, http://hitectransportation.org/news/2009/Exec_Summary02-2009.pdf. E-mail from Todd West of Sandia National Laboratory, February 10, 2010. Estimates vary based on the assumptions made for conversion yields, feedstock availability and relative cost, and more.

h. Biomass Research and Development Initiative, *Increasing Feedstock Production for Biofuels Economic Drivers, Environmental Implications, and the Role of Research*, December 2008, http://www.brdisolutions.com.

i. Agricultural lands include corn stover and straw.

How Many Commercial Cellulosic Biofuel Plants Exist?

As of mid-2010, no large-scale commercial cellulosic biofuel plants are in operation in the United States. Several small-scale plants may come online this year (see **Table 4**). Proposed cellulosic ethanol plants span an array of production capacities—ranging from 20,000 gallons per year to 100 mgy.[36] Development and construction for some commercial cellulosic ethanol plants began in 2008. Industry sources anticipate that many of the plants will be fully operational by 2012. More than two dozen demonstration- or pilot-scale cellulosic ethanol plants are reported to exist currently in the United States.

Table 4. Estimated National Cellulosic Ethanol Plant Tally

Source	Total Number of Commercial Plants / Combined Production Capacity	Total Number of Pilot or Demonstration Plants / Combined Production Capacity	Total Number of Proposed, Planned, or Under Construction Plants / Combined Production Capacity	Comments
Purdue University[a]	0	25 / 3.5 mgy	30 / 1bgy	
GAO[b]	0		25 / 376 mgy	Data for the GAO study was obtained from the Renewable Fuels Association.
Renewable Fuels Association[c]	0	not provided	20 / 242 mgy	Two of the plants may come online in 2010.
				Total planned production capacity is based on estimates provided for initial stages of operation.
EPA[d]	0	23 / 4.1mgy	17 / 265 mgy	

Source: Compiled by CRS.

Notes: mgy = million gallons per year. bgy = billion gallons per year.

a. Wallace E. Tyner and Sarah Brechbill, "Cellulosic Biofuels: Feedstocks, Conversion Technologies, Economics, and Policy Issues," CRS Workshop on the Development of the U.S. Cellulosic Biofuels Industry, Washington, DC, October 6, 2009; and conversation with Wallace Tyner from Purdue University, February 2, 2010.

b. GAO, *Bio fuels: Potential Effects and Challenges of Required Increases in Production and Use*, p. I 18.

c. Renewable Fuels Association, *U.S. Cellulosic Ethanol Projects Under Development and Construction*, February 25, 2010, http://www.ethanol rfa.org/resource/cel I u losic/docu ments/ Cu rrentAdvancedCel I u loseBiofuelsProjects2-25- I 0.pdf.

d. EPA, *Renewable Fuel Standard Program (RF52) Regulatory Impact Analysis*, EPA-420-R-10-006, Washington, DC, February 2010, pp. 171 and 186, *http://www.epa.gov/oms/r enewablefue* Is/420r I 0006.pdf.

WHAT POLICY OPTIONS ARE AVAILABLE TO EXPEDITE CELLULOSIC BIOFUEL PRODUCTION?

Congress is considering legislation that will affect cellulosic biofuel production if enacted. Section 129 of the American Clean Energy and Security Act of 2009 (H.R. 2454, also known as Waxman-Markey) would amend the Loan Guarantee Program to incorporate renewable fuel pipeline construction.[37] Title I of the American Clean Energy Leadership Act of 2009 (ACELA, S. 1462) would amend the Loan Guarantee Program and create a Clean Energy Deployment Administration, under DOE, to advance lending and implementation of commercial clean energy technologies.[38] H.R. 2283 and S. 943 would waive the lifecycle greenhouse gas emission reduction requirements for renewable fuel production.

Congress may decide to take additional legislative action to address the limited cellulosic biofuel production capacity for the RFS. Some options available to Congress are lowering the cellulosic biofuel mandate, delaying the start year for cellulosic biofuel production requirements, modifying the DOE Loan Guarantee Program, implementing new financial support mechanisms, or making federal lands available for biomass removal. Cellulosic biofuels advocates may fmd it beneficial for Congress to make the cellulosic biofuel mandate more attainable in the near term (e.g., two to five years). Opponents may view any additional congressional action to assist the cellulosic biofuel community as harmful to the entire renewable energy market in the long run.

Congress may also choose to monitor EPA's facilitation of the RFS. EPA is responsible for implementing the RFS and revising the RFS when necessary. In May 2009, EPA announced a notice of proposed rulemaking for the RFS seeking public comment for a number of issues related to the RFS, including the cellulosic biofuel production volume, the definition of cellulosic biofuel, and the lifecycle emissions analysis.[39] On March 26, 2010, EPA published its final rule for the RFS.[40] The final rule was effective as of July 1, 2010. As discussed above, the most significant change pertaining to cellulosic biofuels was issuance of a waiver that lowered the 2010 cellulosic biofuel mandate to 6.5 million gallons. Some suggest the waiver was issued partly due to the economic hardships faced by the industry.[41]

End Notes

[1] The RFS was expanded under the Energy Independence and Security Act of 2007 (EISA; P.L. 110-140). For more information on the expanded RFS, see CRS Report R40155, *Renewable Fuel Standard (RFS): Overview and Issues,* by Randy Schnepf and Brent D. Yacobucci

[2] U.S. Environmental Protection Agency, "Regulation of Fuels and Fuel Additives: Changes to Renewable Fuel Standard Program; Final Rule," 75 *Federal Register,* March 26, 2010.

[3] For more information on cellulosic biofuels, see CRS Report RL34738, *Cellulosic Biofuels: Analysis of Policy Issues for Congress,* by Kelsi Bracmort et al.

[4] For more information on ethanol, see CRS Report RL33290, *Fuel Ethanol: Background and Public Policy Issues,* by Brent D. Yacobucci.

[5] Cellulose feedstocks can also be used to provide heat or generate electricity via gasification, combustion, anaerobic digestion, and other conversion processes. For more information on anaerobic digestion, see CRS Report R40667, *Anaerobic Digestion: Greenhouse Gas Emission Reduction and Energy Generation,* by Kelsi Bracmort.

[6] R. M. Cruse and C .G. Herndl, "Balancing Corn Stover Harvest for Biofuels with Soil and Water Conservation," *Journal of Soil and Water Conservation,* vol. 64, no. 4 (July/August 2009), pp. 286-291.

[7] For more information on the RFS, see CRS Report R40155, *Renewable Fuel Standard (RFS): Overview and Issues,* by Randy Schnepf and Brent D. Yacobucci.

[8] U.S. Dept. of Energy, U.S. Dept of Agriculture, *Biomass as Feedstock for a Bioenergy and Bioproducts Industry: The Technical Feasibility of a Billion-Ton Annual Supply,* April 2005, http://wwwl.eere.energy final_billionton_vision_report2.pdf.

[9] Diane Greer, "Creating Cellulosic Ethanol: Spinning Straw into Fuel," *BioCycle,* April 2005; Biotechnology Industry Organization, *Achieving Sustainable Production of Agricultural Biomass for Biorefinery Feedstock,* Washington, DC, 2006, http://www.bio.org/ind/biofuel and Biotechnology Industry Organization, "Energy Bill Biofuels Mandates Will Be Achievable with Biotechnology Advances," press release, November 18, 2007, http://bio.org/news/pressreleases/newsitem.asp?id=2007_1218_01&p=yes.

[10] Ian Talley, "Renewed Energy: US Biofuel Mandate Calls for Big Production Boost," *Dow Jones International News,* December 18, 2007, at http://www.factiva.com/.

[11] For more information on EPA's waiver authority, see CRS Report RS22870, *Waiver Authority Under the Renewable Fuel Standard (RFS),* by Brent D. Yacobucci.

[12] 42 U.S.C. 7545(o)(7).

[13] EPA, "Changes to Renewable Fuel Standard; Final Rule," p. 14.

[14] U.S. Environmental Protection Agency, "Regulation of Fuels and Fuel Additives: 2011 Renewable Fuel Standards; Proposed Rule," 75 *Federal Register* 42242, July 20, 2010.

[15] The original RFS established by § 1501 of EPAct05 required 4.0 billion gallons of renewable fuel for 2006, ascending to 7.5 billion gallons by 2012. The original RFS would have required that 250 million gallons of the renewable fuel be derived from cellulosic biomass starting in 2013. The current RFS, established under EISA, had required 250 million gallons of cellulosic biofuel starting in 2011. However, as discussed above, the EPA recently announced it would lower the 2011 mandate to somewhere between 5 and 17.1 million gallons.

[16] Clean Fuels Development Corporation, Nebraska Ethanol Board, and U.S. Dept of Agriculture, *A Guide for Evaluating the Requirements of Ethanol Plants,* 2006, http://www.ethanol guide_for_evaluatingtherequirementsof ethanol_plants.pdf.

[17] For more information on the blender tax credit, see CRS Report RL34130, *Renewable Energy Programs in the 2008 Farm Bill,* by Megan Stubbs

[18] A loan guarantee is defined as a "pledge with respect to the payment of all or a part of the principal or interest on any debt obligation of a non-federal borrower to a non-federal lender." The LGP was first authorized under Title XVII of EPAct05 and then amended

under the American Recovery and Reinvestment Act of 2009 (P.L. 111-5). DOE may issue loan guarantees to eligible projects that "avoid, reduce, or sequester air pollutants or anthropogenic emissions of greenhouse gases" and "employ new or significantly improved technologies as compared to technologies in service in the United States at the time the guarantee is issued." Eligible projects include commercial-scale renewable energy systems. EISA authorized the DOE to issue loan guarantees in part to support renewable energy projects.

[19] DOE considers a project commercial-scale if it converts, at a minimum, 700 tons of biomass per day to energy. DanTobin, "Biomass Summit," *The DOE Loan Guarantee Program: A Status Report,* Washington, DC, October 20, 2009.

[20] For more information on appropriations for the LGP, see CRS Report R40669, *Energy and Water Development: FY2010 Appropriations,* coordinated by Carl E. Behrens.

[21] The funding amount listed in ARRA is $6 billion to cover certain costs, but this amount was recently reduced to $4 billion, as $2 billion was transferred to the "cash for clunkers" automobile trade-in program. For more information on appropriations for DOE loan guarantees and direct loans, see CRS Report R40669, *Energy and Water Development: FY2010 Appropriations,* coordinated by Carl E. Behrens; and CRS Report RL30346, *Federal Credit Reform: Implementation of the Changed Budgetary Treatment of Direct Loans and Loan Guarantees,* by James M. Bickley.

[22] Renewable Fuels Association, "2010 State of the Industry Address," 2010 National Ethanol Conference, Orlando, FL, February 16, 2010, *http://www.ethanolrfa.org/objects/documents/2772/2010_state* the_industry.pdf? utm_medium=email&utm_source=Emailmarketingsoftware&utm_content=313532555&utm_campaign= 2010S0T1+_+osdik&utm_term=StateoftheIndustryaddress.

[23] U.S. Dept. of Energy, "Secretaries Chu and Vilsack Announce More Than $600 Million Investment in Advanced Biorefinery Projects," press release, December 4, 2009, http://wwwl.eere.energy news_id=15660.

[24] BCAP receives its authorization from Title IX of the Farm Security and Rural Investment Act of 2002 (P.L. 107-171) and was amended by Title IX of the Food, Conservation, and Energy Act of 2008 (P.L. 110-246). For more information on BCAP, see CRS Report R41296, *Biomass Crop Assistance Program (BCAP): Status and Issues,* by Megan Stubbs, or visit http://www.fsa.usda.gov/FSA/webapp?area=home&subject=ener&topic=bcap.

[25] Juliet Eilperin, "The Unintended Ripples from the Biomass Subsidy Program," *Washington Post,* January 10, 2010, p. A3; and Nathanael Greene, Natural Resources Defense Council Switchboard, "USDA Sets BCAP on Horrid Path," blog post, January 26, 2010, http://switchboard.nrdc.orgiblogs/ngreene/usda_sets_bcap_on_horrid_path.html.

[26] For more information on biomass definitions, see CRS Report R40529, *Biomass: Comparison of Definitions in Legislation,* by Kelsi Bracmort and Ross W. Gorte

[27] University of Tennessee, *Growing and Harvesting Switchgrass for Ethanol Production in Tennessee,* SP701-A, http://www.utextension.utk.edu/publications/spfiles/SP701-A.pdf.

[28] The plant is scheduled to come online in mid-2010 with an expected utilization rate of 50%. U.S. Environmental Protection Agency, *Renewable Fuel Standard Program (RFS2) Regulatory Impact Analysis,* EPA-420-R-10-006, Washington, DC, February 2010, p. 175, http://www.epa.gov/oms/renewablefuels/420r10006.pdf.

[29] Renewable Fuels Association, *U.S. Cellulosic Ethanol Projects Under Development and Construction,* February 25, 2010, *http://www.ethanolrfa.org/resource/cellulosic/documents/CurrentAdvancedCelluloseBiofuelsProjects2-25-* 10.pdf. Projects were selected based on operation dates and production capacity for first stage of operation provided. Projects identified as small-scale or pilot-scale were not included in the tally.

[30] Based on information provided for 20 U.S. cellulosic ethanol projects under development and construction by the Renewable Fuels Association, *http://www.ethanolrfa.org/resource/cellulosic/documents/* CurrentAdvancedCelluloseBiofuelsProjects2-25-10.pdf.

[31] Dan Chapman, "Bio Energy Backers Stay Upbeat Despite Setbacks," *Atlanta Journal-Constitution*, January 13, 2010.

[32] Russel Gold and Siobhan Hughes, "Biofuel Production Fall Far Short of Targets," *Wall Street Journal,* February 4, 2010.

[33] Renewable Fuels Association, "Obama Administration on Right Track with Biofuels," press release, February 3, 2010, http://renewablefuelsassociation. email 1 .com/TNiewE mail/y/9C2814171DC11FF6/ F83B91963160BD91C5ECO8CADFFC107B.

[34] A biorefinery is a facility that processes biomass into biofuels.

[35] The estimates provided in **Table 3** are based on numerous assumptions and modeling techniques unique to each study. While general categories may appear similar in name (e.g., agricultural lands), one should refer to the table footnotes for clarification on the category makeup.

[36] Renewable Fuels Association, *U.S. Cellulosic Ethanol Projects Under Development and Construction,* February 25, 2010, *http://www.ethanolrfa.org/resource/cellulosic/documents/CurrentAdvancedCelluloseBiofuelsProjects2-25-* 10.pdf.

[37] For more information on expansion of the Loan Guarantee Program, see CRS Report R40643, *Greenhouse Gas Legislation: Summary and Analysis of H.R. 2454 as Passed by the House of Representatives*, coordinated by Mark Holt and Gene Whitney

[38] For more information on measures that may improve cellulosic ethanol production contained within S. 1462, see CRS Report R40837, *Summary and Analysis of S. 1462: American Clean Energy Leadership Act of 2009, As Reported,* coordinated by Mark Holt and Gene Whitney.

[39] U.S. Environmental Protection Agency, "Regulation of Fuels and Fuel Additives: Changes to Renewable Fuel Standard Program," 74 *Fed. Reg.* 24904-25143, May 26, 2009.

[40] U.S. Environmental Protection Agency, "Regulation of Fuels and Fuel Additives: Changes to Renewable Fuel Standard Program; Final Rule," *75 Federal Register,* March 26, 2010.

[41] Renewable Fuels Association, January 2010, personal consultation.

In: Cellulosic Biofuels: Feedstock, Technology… ISBN: 978-1-61209-509-7
Editors: Robert M. Burkin et al.

Chapter 4

RESEARCH ADVANCES: NREL LEADS THE WAY - CELLULOSIC ETHANOL

National Renewable Energy Laboratory

FUELING THE FUTURE

On the Road to Energy Security

One of our greatest challenges is to reduce our nation's dependence on imported petroleum. To accomplish this, we need a variety of alternative fuels, including ethanol produced from cellulosic materials like grasses and wood chips. Fortunately, the United States has abundant agricultural and forest resources that can be converted into biofuels. Recent studies by the U.S. Department of Energy (DOE) suggest these resources can be used to produce enough ethanol – 60 billion gallons/year – to displace about 30% of our current gasoline consumption by 2030.

How do we get there? Currently, there are no commercial cellulosic ethanol refineries. The ethanol we use is derived primarily from corn kernels, a form of starchy biomass. When manufacturers produce ethanol from corn, they use enzymes to convert starches to simple sugars and yeasts to ferment the sugars into ethanol. Cellulosic biomass contains sugars as well, but they are much harder to release than those in starchy biomass. To complicate matters, the process of releasing the sugars produces by-products that inhibit fermentation, and some of the sugars from cellulosic biomass are difficult to ferment.

All this makes cellulosic ethanol production complicated—and expensive. To displace petroleum, cellulosic ethanol must be cost competitive. DOE has determined that competitiveness can be achieved at an ethanol production cost of $1.07/gallon (in 2002 dollars) and aims to achieve this goal by 2012. To do this, the technology used to produce cellulosic ethanol must be improved. That's where the National Renewable Energy Laboratory (NREL) comes in.

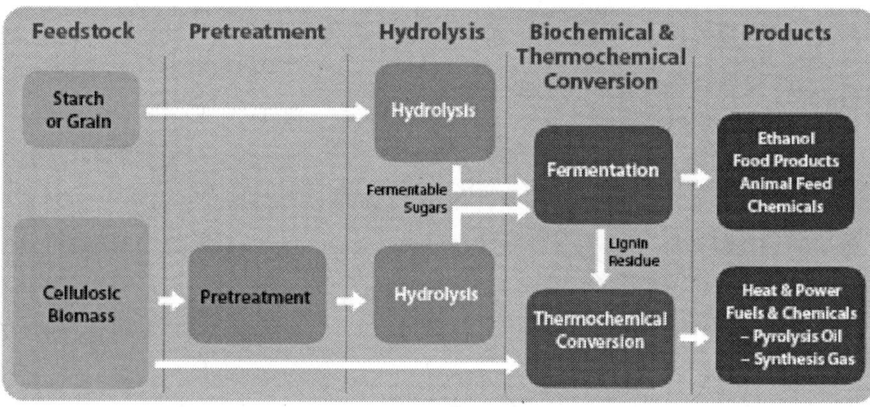

Schematic of a hypothetical integrated ethanol biorefinery

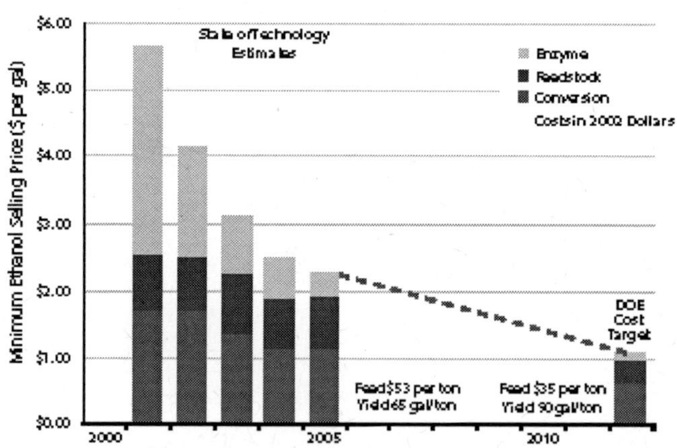

NREL has led progress toward DOE's cost target of $1.07/gallon for biochemically produced ethanol. Similar progress is being made for thermochemically produced ethanol.

NREL leads DOE's National Bioenergy Center and is on the cutting edge of cellulosic ethanol technology. NREL's research addresses each step of the processes that produce cellulosic ethanol and valuable co-products. NREL's research covers the full spectrum from fundamental science and discovery to demonstration in fully integrated pilot plants.

This brochure highlights some of NREL's recent advances in cellulosic ethanol production. Research at NREL addresses both biochemical (chemicals, enzymes, and fermentative microorganisms) and thermochemical (heat and chemical) processes. For the biochemical processes, NREL investigates pretreatment, hydrolysis, and fermentation steps as well as process integration and biomass analysis. For the thermochemical processes, NREL researches catalyst development, process development, and process analysis.

PRETREATMENT

Improving the Critical First Step toward Cost-Competitive Ethanol

To break down cellulose—the primary source of sugar in fibrous biomass—you have to first get past hemicellulose and lignin, which surround the cellulose in a protective sheath. This is the job of pretreatment. NREL typically uses a moderately high-temperature, high-pressure dilute acid pretreatment process to break down (hydrolyze) hemicellulose and disrupt or dissolve lignin. Hydrolyzing the hemicellulose also creates another important source of soluble sugars for later fermentation to ethanol.

NREL is investigating potentially cheaper, but still effective, pretreatment methods. In one recent advance, NREL researchers applied their knowledge of biomass structural changes to pretreatment process development. Lignin dissolved under certain pretreatment conditions can apparently redeposit onto cellulose, creating a barrier to effective cellulose hydrolysis and reducing sugar yield. NREL is using its state-of-the-art imaging and analytical tools to understand lignin redeposition and design pretreatment processes that minimize its detrimental effects.

In another recent advance, NREL employed enzymes to enable milder pretreatment. Although dilute acid pretreatment can break down hemicellulose very effectively, the severe conditions require expensive processing equipment and tend to degrade the sugars. Using a milder pretreatment process could cut process costs dramatically and eliminate sugar degradation losses. The challenge is to maintain a high level of effectiveness with the milder process, which is accomplished by using enzymes to further break down the hemicellulose after pretreatment. NREL has shown that proper mixtures of enzymes can enhance hemicellulose hydrolysis. In an experiment on pretreated corn stover, adding a hemicellulase enzyme to break down the hemicellulose increased the yield of xylose (a sugar resulting from hemicellulose hydrolysis) by 12% across a range of pretreatment conditions. Breaking down the hemicellulose also enhanced cellulose hydrolysis, resulting in a 6% higher glucose yield.

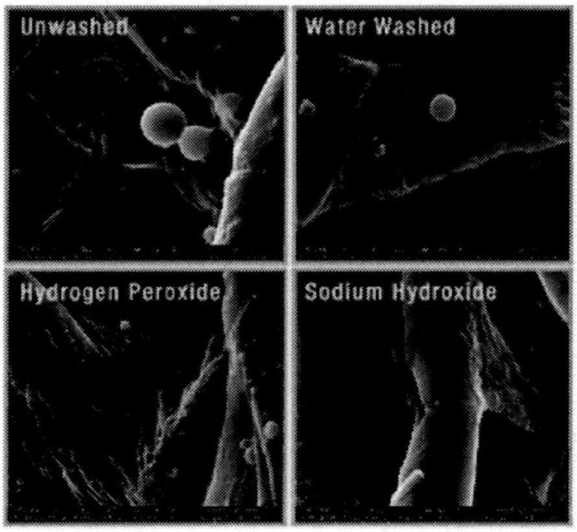

NREL used scanning electron microscopy to reveal what are thought to be lignin droplets remaining on pretreated filter paper after washing with various solvents. (T. Vinzant, NREL)

This pretreatment reactor hydrolyzes hemicellulose and solubilizes some lignin. The pressurized hot wash process separates these materials before they can reprecipitate. This system uses dilute sulfuric acid at increased temperature and pressure, but pressurized hot wash may work well with any pretreatment system.

To get a broader look at pretreatment options, NREL participates in the Biomass Refining Consortium for Applied Fundamentals and Innovation (CAFI). Each CAFI participant is evaluating a different pretreatment approach using standardized experimental design and data reporting protocols. The CAFI projects allow participants to compare pretreatment and downstream process performance across a range of lignocellulo sic feedstocks. NREL is using this knowledge to help identify the best pretreatment approaches for near- and long-term biorefining platforms. (For more information on CAFI, see "Coordinated Development of Leading Biomass Pretreatment Technologies," *Bioresource Technology*, December 2005.)

ENZYMATIC HYDROLYSIS

Unlocking the Full Potential of Cellulosic Biomass

Plants have evolved over several hundred million years to be recalcitrant—resistant to attacks from the likes of bacteria, fungi, insects, and extreme weather. Breaking down plants is no easy task. For cellulosic ethanol production, the primary challenge is breaking down (hydrolyzing) cellulose into its component sugars.

NREL is exploring the causes of biomass recalcitrance and ways to overcome it using cellulases (enzymes that break down cellulose). The goals are to maximize the conversion of cellulose to sugar, accelerate the rate of conversion, and use fewer, cheaper enzymes. NREL's recent advances include employing state-of-the-art capabilities to characterize plant structure and developing superior enzymatic hydrolysis processes.

To make contact with cellulose, the enzymes must get past a complex maze of plant structures. NREL is mapping this labyrinth as a first step toward overcoming it. A unique array of microscopy tools and techniques in NREL's new Biomass Surface Characterization Laboratory enables researchers to image plant structures down to the molecular level. To probe even further—visualizing structures and processes at scales that cannot (yet) be observed—NREL and its partners are building a sophisticated molecular dynamics model of the cellulose-cellulase system. When complete, it will be the largest biological computer model ever developed.

Advanced computer modeling capabilities help NREL understand and improve enzymatic hydrolysis processes. This illustration of the CBH I enzyme is based on models developed by NREL.

Once cellulases make contact with cellulose, the real work begins. Cellulases act very slowly. That's why dead trees take years to decompose in the forest. To accelerate cellulose conversion, it is critical to start with the best enzymes nature has to offer. The most active known cellulases are in the cellobiohydrolase I (CBH I) family, derived from fungi. But not all CBH I enzymes are equal. NREL recently confirmed the existence of CBH I enzymes that are twice as active as those from industrial sources.

NREL and its partners Genencor International and Novozymes have developed a "cocktail" of cellulases to improve hydrolysis. In combination with NREL's process development improvements, this advance has reduced enzyme cost twentyfold. This work received an R&D 100 Award in 2004.

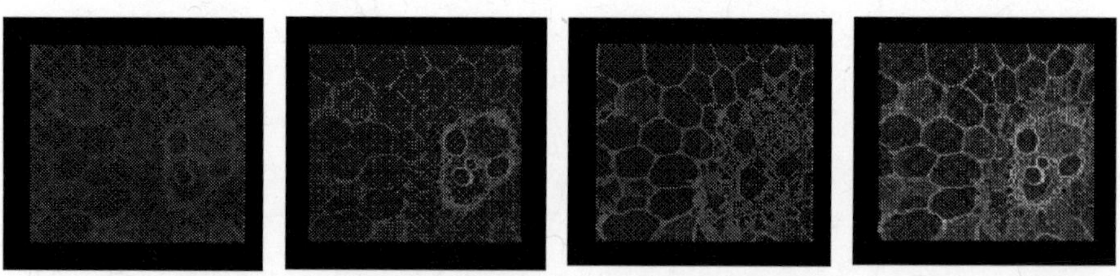

These ultra-sharp laser microscope images were created with the Biomass Surface Characterization Laboratory's scanning confocal microscope, which can be used to build 3-D representations of plant structures. (S.Porter, NREL)

FERMENTATION

Creating "Super-bugs" for Superior Ethanol Yield

During fermentation, microorganisms (primarily fungi and bacteria) convert the sugars in biomass to ethanol. Under ideal conditions, these "bugs" will work contentedly, consuming sugars and producing ethanol and other products. But conditions in a cellulosic ethanol biorefinery are anything but ideal.

The hot soup—called a hydrolyzate—generated after pretreatment and hydrolysis contains not only fermentable sugars, but also compounds (such as acetic acid) that are toxic to the bugs. Other things that are toxic in the fermentation process and the hydrolyzate are a high-solids concentration and a rising ethanol concentration. Because microorganisms found in nature do not function well in this hostile environment, NREL is creating "super-bugs" that thrive in it.

Yeasts are currently the fermentation organisms of choice for the corn ethanol industry. They are reasonably tolerant of ethanol, acid, and moderately high temperatures. However, existing yeast strains cannot withstand highly toxic hydrolyzates or ferment 5- carbon sugars and minor 6-carbon sugars efficiently. NREL, along with the National Corn Growers Association (NCGA) and Corn Refiners Association (CRA), developed yeast capable of fermenting a particular 5-carbon sugar, arabinose, which constitutes up to 20% of the fermentable sugars in corn fiber. Three genes from a bacterium were inserted into the yeast Saccharomyces cerevisiae. This work resulted in the first ever demonstration, in 2000, of arabinose fermentation by yeast. Next, NREL plans to test the strain under real biorefining conditions—in the hydrolyzate.

NREL also pioneered the use of a yeast alternative, the bacterium Zymomonas mobilis (Zymo). Zymo gives a high ethanol yield and tolerates high ethanol concentrations. Using genetic and metabolic engineering, NREL developed acetic acid-tolerant Zymo strains that can ferment arabinose and the most important 5-carbon sugar, xylose. This strain resulted in several patents and an R&D 100 Award. NREL also pioneered a technique to make the Zymo strain stable (the bacteria's offspring have the same genes as the parents) by inserting key genes into the genome. NREL's Zymo work has included successful collaborations with the NCGA and CRA, the chemical company Arkenol (now BlueFire Ethanol), and DuPont.

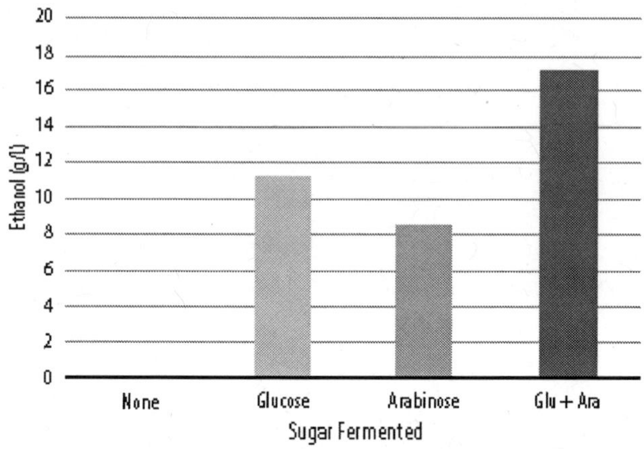

The yeast strain developed by NREL, NCGA, and CRA is the world's first yeast to ferment arabinose. These results show an ethanol yield of 83% from arabinose in a defined medium (not a hydrolyzate). From left to right, initial sugar concentrations were 0 g/L, 20 g/L glucose, 20 g/L arabinose, and 20 g/L glucose + 20 g/L arabinose. Expected ethanol from 20 g/L of sugar is 10.2 g/L at 100% yield. (A. Singh, NREL, patent pending)

Warren Gretz, NREL PIX 00945

Fermentation processes are tested in NREL's biochemical process development unit (PDU). This 9,000-L fermenter is large enough to produce sufficient lignin for processing in the downstream thermochemical PDU.

PROCESS INTEGRATION

Tying together the Integrated Biorefinery

To produce low-cost ethanol, biorefineries will need to link the refining steps into an integrated process. However, optimizing conditions in one step of the process can influence performance in other steps. The challenge is to find the right combination of trade-offs that optimize the integrated process. Studying integrated biorefinery operations requires an advanced process development unit and state-of-the-art chemical analysis capabilities. NREL's research on high-solids operation is a good example.

NREL has shown that high-solids operation—using a high ratio of biomass to water in the biorefining process—is one key to cutting ethanol costs. The less water introduced in the pretreatment step, the higher the potential sugar concentration and the less equipment and energy the process requires. The result is lower-cost ethanol. In a perfect world that would be enough, but high solids concentrations can create problems elsewhere in the process.

In a first-of-its-kind study, NREL demonstrated that a moderately high solids concentration combined with recycled process water severely inhibits fermentation and, consequently, lowers ethanol yield. This is important because commercial biorefineries will need to recycle water, taking it from the back end of the refining process and combining it with fresh water at the front end of the process. NREL's study identified the ability to achieve high solids concentration and high levels of process water recycle as an issue that must be considered in both pretreatment and fermentative microorganism development.

Managing the properties of high-solids mixtures is another process integration challenge. Like adding flour to water in a recipe, adding biomass to water makes the mixture thicker and more viscous. This has important implications for the efficient flow and conversion of biomass through the integrated process. NREL is developing unique capabilities in biomass rheology—the science of the deformation and flow of materials—to determine the best ways to manage high-solids biomass mixtures. This research is critical to providing process engineers with rheological information needed to design a commercial biorefinery.

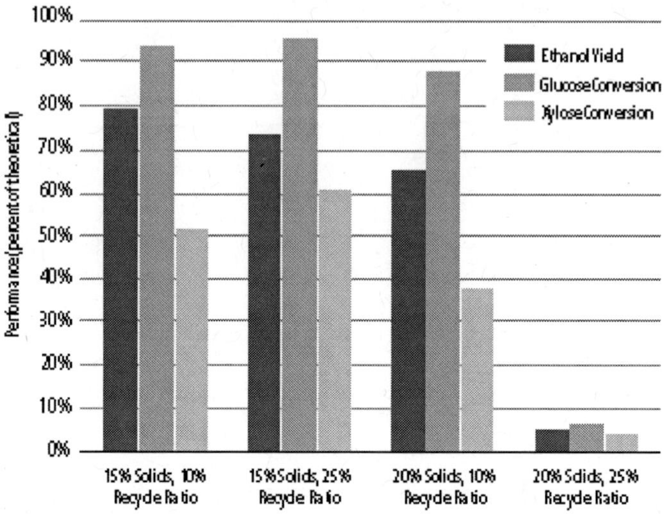

Pioneering NREL research shows ethanol yield dropping dramatically at moderately high solids concentration and high recycle ratio (the ratio of recycled water to fresh water). (D. Schell, NREL)

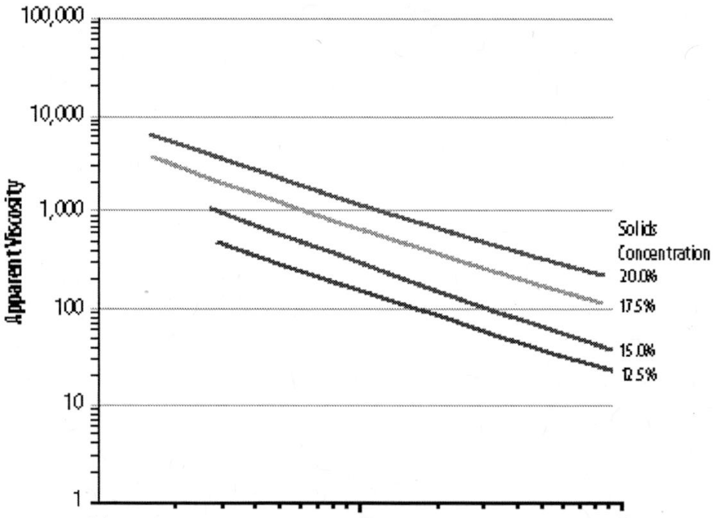

NREL is quantifying the rheological properties of various biomass materials. These results show how the viscosity (resistance to flow) of pretreated corn stover increases with higher solids concentration and decreases with faster shear rate (speed of mixing). (J. McMillan, NREL)

THERMOCHEMICAL CONVERSION

Honing a Powerful Path to Economic Ethanol

The advances described up to this point relate to biochemical conversion of carbohydrates to ethanol. However, ethanol can also be produced thermochemically from any form of biomass. In this approach, heat and chemicals are used to break biomass into syngas (CO and H2) and reassemble it into products such as ethanol. This method is particularly important because up to one third of cellulosic biomass—the lignin-rich parts—cannot be easily converted biochemically.

Forest products and mill residues typically have high lignin contents, making them unattractive feedstocks for biochemical conversion yet suitable for thermochemical conversion. In an integrated biorefinery, lignin-rich residues from the biochemical process could also be converted thermochemically.

A thermochemical biomass conversion process is complex, and uses components, configurations, and operating conditions that are more typical of petroleum refining. And, just like researchers in the petroleum industry, NREL uses a combination of experimental research together with process economic models to explore a large number of possible process configurations. A much simplified schematic of NREL's preferred configuration is shown below. This configuration employs an indirect gasifier, tar reforming, and a mixed alcohol synthesis step designed to maximize ethanol yield.

In thermochemical conversion, biomass is converted into syngas, and syngas is converted into an ethanol- rich mixture. However, syngas created from biomass is not "clean"—it contains contaminants such as tar and sulfur that interfere with the conversion of the syngas into products. These contaminants must be removed. NREL has developed tar-reforming catalysts and catalytic reforming processes that have demonstrated high levels of tar conversion— converting up to 97% of the tar into more syngas. This not only cleans the syngas, it also creates more of it, improving process economics and ultimately cutting the cost of the resulting ethanol. NREL has also made progress regenerating the tar-reforming catalyst after it has been partially deactivated by sulfur poisoning.

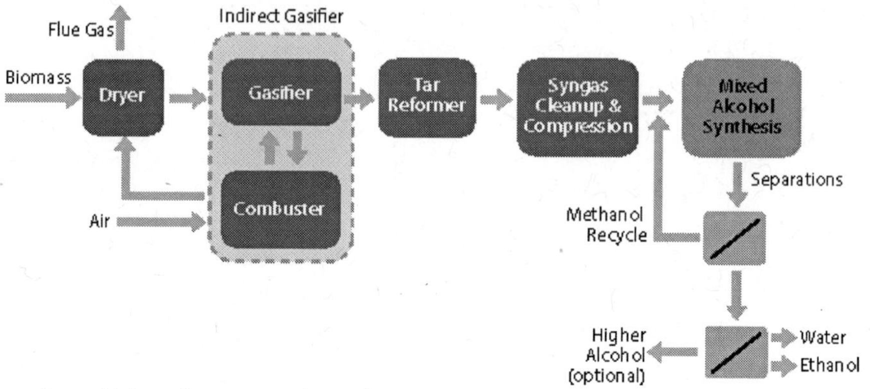

This simplified schematic of NREL's thermochemical conversion model shows the many steps that can be manipulated to optimize efficiency and cost.

NREL is evaluating many different process options and their associated costs to help identify key barriers to low-cost ethanol production. For example, process models indicate that reducing tars and hydrocarbons from syngas can decrease the production cost of ethanol by 33%. NREL models also highlight the need for extensive heat integration and quantify performance targets needed to achieve DOE's ethanol cost goals thermochemically.

BIOMASS ANALYSIS

Enabling Total Control of Ethanol Production

You can't control what you can't measure. Biorefinery operators will need to know the precise composition of the biomass going through their processes so they can tightly control the cost and quality of the products coming out. The faster and more reliable the measurements are, the better and cheaper the final products will be.

NREL's capabilities are constantly evolving to meet industry's need for accurate and rapid biomass analysis. Researchers recently developed a way to rapidly analyze biomass composition using near- infrared (NIR) spectroscopy. In this technique, light reflected off a biomass sample is analyzed to determine the sample's composition. Compared with traditional wet chemistry analysis, this is a huge leap forward. Analyzing a sample with this new approach takes minutes instead of weeks and costs tens of dollars instead of thousands of dollars—without sacrificing precision or accuracy. The rugged NIR instruments can even be adapted to a working biorefinery to measure, for example, the chemical composition of corn stover as it enters and exits pretreatment.

What if you could measure an adult plant's composition while it is still a sprout? You could grow plants that yielded the most ethanol, maximize cellulose content, and minimize lignin. Or what if, by knowing how genes affect plant composition, you could create the ideal ethanol feedstock? NREL's analytical capabilities are making these scenarios a reality. Using techniques such as molecular beam mass spectrometry and nuclear magnetic resonance spectroscopy, NREL is measuring cell wall chemistry more quickly and accurately than anyone else in the world. These measurements are used to predict adult plant composition. In 2006, NREL analyzed the cell wall chemistry of more than 10,000 samples for industry and university partners. These thermoanalytical capabilities can help accelerate crop breeding and genetic engineering.

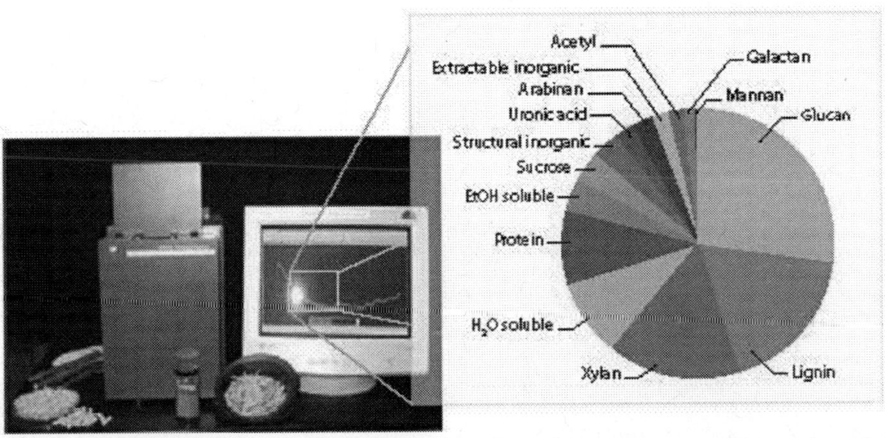

NREL's NIR spectroscopy technique rapidly analyzes biomass composition and can be adapted to on-line use.

Chapter Sources

The following chapters have been previously published:

Chapter 1 – This is an edited, excerpted and augmented edition of a United States Congressional Research Service publication, Report Order Code R41460, dated October 22, 2010.

Chapter 2 – This is an edited, excerpted and augmented edition of a United States Congressional Research Service publication, Report Order Code RL34738, dated October 14, 2010.

Chapter 3 – This is an edited, excerpted and augmented edition of a United States Congressional Research Service publication, Report Order Code R41106, dated July 28, 2010.

Chapter 4 – This is an edited, excerpted and augmented edition of a National Renewable Energy Laboratory publication, dated March 2007.

INDEX

C

F

G

M

N

O

T

U

V

W

Y